Radio, Television, and Sound System Repair

An Introduction

JOEL GOLDBERG

Prentice-Hall, Inc., *Englewood Cliffs, New Jersey* 07632

Library of Congress Cataloging in Publication Data

GOLDBERG, JOEL, (date)
   Radio, television, and sound system repair.

   Includes index.
   1. Radio—Repairing. 2. Television—Repairing.
3. Sound—Recording and reproducing—Equipment and
supplies—Maintenance and repair. I. Title.
TK6553.G576   621.38   77-20968
ISBN 0-13-752238-X

Printed in the United States of America

10 9 8 7 6 5 4 3

PRENTICE-HALL INTERNATIONAL, INC., *London*
PRENTICE-HALL OF AUSTRALIA PTY. LIMITED, *Sydney*
PRENTICE-HALL OF CANADA, LTD., *Toronto*
PRENTICE-HALL OF INDIA PRIVATE LIMITED, *New Delhi*
PRENTICE-HALL OF JAPAN, INC., *Tokyo*
PRENTICE-HALL OF SOUTHEAST ASIA PTE. LTD., *Singapore*
WHITEHALL BOOKS LIMITED, *Wellington, New Zealand*

Contents

Preface

This book differs from most books on electronics in that the author looks at the *why* of electronics before he discusses the specific theoretical aspects. Traditionally, books on electronics present, in sequence, direct-current concepts, alternating-current concepts, vacuum tubes and transistors, and then amplifiers, oscillators, and rectifiers, before looking at complete units and the *why* of these units. Students enrolled in traditional electronics-programs spend many hours in the laboratory, building and testing resistive circuits designed to represent loads in electronics. In many instances, the theory trees are so thick that the student fails to see the forest. The student often has a difficult time in accepting the concept of the resistor's representing a functioning load. The student also has difficulty in accepting the principle that the vacuum tube and solid-state device are in actuality controlled resistances in the circuit. This failure of understanding makes it difficult for the student to determine component failure and retards his ability to develop speedy troubleshooting techniques.

The author believes that a student who knows how an electronic device functions, and is able to use a manufacturer's technical literature as well as testing equipment to measure operating and signal voltages, should be highly successful as a repair technician. In order to motivate the student early in the program, the author first looks at input and output devices as used in home-entertainment electronic units. The next phase of the material is a discussion of functional block diagrams for typical devices. Thirdly, signal paths are investigated, followed by a presentation of theoretical material related to vacuum tubes and semiconductors. The fundamentals of troubleshooting malfunctions in basic devices are dealt with last. Throughout the book, electrical theories are presented when they are relevant to the material being covered. This emphasis on theory when required makes comprehension of the material seem a natural occurrence and promotes high student interest. The student is able to relate the learned concepts readily to circuit applications, thus enhancing his ability to perform successfully on the job.

The author's intent is that this book will be used in senior high schools and adult education classes, as well as in technical schools and colleges, as a basic text. The vocabulary has been carefully chosen so that the reader will be able to comprehend the material with little difficulty.

A venture as complex as the authorship of a textbook often becomes a group effort due to the many investigations and discussions occurring prior to and during the actual writing. Many others have contributed to the development of this book. I wish to thank my colleagues Robert Critchett and E. Eugene Ranta for their comments and suggestions during the preparation and writing of this book.

I am especially grateful to my wife, Alice, for her hours of assistance with the editing and typing of the manuscript, and to our children Michael and Linda for their reactions and suggestions as the book was being written. Their encouragement and compassion did much to help bring this project to its completion.

JOEL GOLDBERG

Section One

Introduction

Introduction

Try, for a moment, to picture a world without a radio, a television, or a tape recorder. It is difficult to envision such a world. Many of us have grown up with a radio in literally every room in our homes. Cars are purchased with factory-installed radios, both AM and FM. Technical advances in the past few years have brought forth a great wave of tape recorders to further entertain us. Compare this picture to one that may have been taken in the 1920's. Radio was a novelty only available to the experimenter. Television was a laboratory device found only in research centers and not available to the public. Tape recorders were virtually unknown.

Today there are over 550 million electronic consumer-products in use for entertainment purposes. Recent sales figures indicate that over 100 million units are being sold yearly. With a population of over 200 million people, approximately 50 million families, a little quick math indicates that the average family owns 11 different electronic entertainment products. Not only does the average family already own this many, but, on the average, it is adding, or replacing, two units per year! These astounding figures help to point out the need for qualified persons who are trained to repair all of the electronic entertainment-units found in the home or car.

The Electronic Industries Association, a group representing about 85 percent of the electronics manufacturers in the United States, has indicated a current need for at least 30,000 qualified technicians in this country. They also project the need for additional personnel as more units are produced and marketed.

In years past, a qualified repair technician could often make a repair simply by testing and changing the vacuum tubes in the radio or television set. This is no longer true. Many of today's sets use transistors and integrated circuits in place of tubes. Repair techniques have changed with these *solid-state* sets.

A successful repair technician must be able to diagnose the faults in the set. First, he must determine in which section of the set the trouble is located. Then, based upon knowledge of *how* the total unit functions as well as what is malfunctioning in the unit, a repair may be effected. The successful technician is able to rapidly repair many sets by applying standard procedures as a functional process in the course of a day's work. Success may be measured in many ways. To the repair technician, the two prime measures of success are the numbers of units properly repaired and the reflection of this in the form of pay for the repairs. Ordinarily, the greater number of repairs is reflected by a larger pay check.

This book is basically a study of the electron in action. The technician is concerned about how the unit functions. In order to determine this, the forces which cause the electron to perform work must be studied. An electron is a part of an atom and it may be controlled in order to perform some useful function. The study of electricity is often considered to be a study of electrons in large quantities. Electronics, on the other hand, is often thought of as the study of the electron as it is used in small quantities. These smaller quantities are typically found in some home entertainment devices using vacuum tubes, transistors, and integrated circuits. Electronics as defined by this author is the study of the electron performing work, as applied to transistors, vacuum tubes, integrated circuits, and associated units employing these components.

This book has been developed in a manner that presents the material in a logical sequence. One of the major concerns of the author is to keep student interest at a high level as the material is covered. For this reason, a *systems* approach is utilized. The author believes that once the student is able to see how the electronic unit (television, radio, etc.) is developed, and can relate this development to the building blocks of the functional unit, then comprehension of the material is easy. Interest in the subject is at a high level, and so there is a recognized need to understand the material.

The information required to effect successful repairs is presented with this systems approach in mind. Most electronic entertainment-devices can be easily repaired when the repair technician has a good understanding of five basic ideas. These five ideas are:

1. The *block diagram* of the unit to be repaired. Understanding the function or purpose of each block in the unit or set is also necessary.

2. *Signal development.* How the signal from the radio, TV station, tape recorder, or phonograph appears in a visual form on an ossciloscope, and the purpose of the various parts that are required to produce the total signal form must be understood.
3. *Signal reception.* The technician must know how the signal used by each of the units is accepted or received by the unit, in other words, what kind of input is used.
4. *Signal processing.* Each unit, after receiving the electrical signal, processes it through one or more functional blocks until the signal reaches the output device. What happens to the signal as it passes through the unit must be understood.
5. *Signal output.* The electrical signal has to be transformed into something the mind is capable of understanding. This requires either a visual or an aural output, such as from a television picture tube or a loudspeaker. The concepts involved in accomplishing this must be understood by the technician if he is to be successful in his repairs and his occupation.

Knowledge of how the system or unit works is necessary before it can be repaired successfully and quickly. It is also necessary for the technician to understand how the various electrical signals, or pieces of information, are processed inside the unit. This book starts by studying input and output devices. Then, using typical home-entertainment units as models, block diagrams of each of these models are reviewed, with an explanation of the function of each of the various blocks that make up typical units. Next, the electrical signal, or information, is traced through each unit in order to help the student understand what happens inside of each audio amplifier, radio, or television set reviewed. A study of electrical theory as it applies to each of the blocks and of the individual components making up each block follows, as a logical step in the mastery of how each of the devices works.

The final step is a study of the methods of troubleshooting. These methods, which are commonly used for most technical or electrical devices, make the repair a logical and often rapid job. Troubleshooting is what the repair field is all about. Simply, it means (1) take the job in; (2) localize the problem to a functional block in the unit; (3) determine which component has failed; (4) replace the defective component; (5) check the unit to make sure the repair was correct; and (6) return the unit to the user. Persons capable of success in performing these steps will meet the needs of industry for knowledgeable technicians able to maintain the many electronic entertainment-devices found in the home or car.

Input
and
Output
Devices

2

In their functioning, all entertainment electronic devices have two major things in common. These two things are that each unit requires some sort of input information and that each emits some sort of output information. This information is provided in many different forms. It may be very simple, such as the single tone produced by a tuning fork; it may be as complex as a color television signal; or, this information can range in complexity somewhere between these two extremes.

Regardless of the complexity of this information, the audio amplifier, radio, tape player, or television set must have some means of taking in, or receiving it, and converting it into an electrical wave or signal. Actually, all the units under study process an electrical signal from some outside source and translate this signal, ultimately, into an intelligible form —either visual or aural, or in some cases, both. The devices used to change information from one form of energy into another form of energy are called *transducers*. For the purposes of this book, transducers are divided into two groups. The first group, classified as *input transducers*, receive the signal from its source and convert it into electrical energy in order to process it through the unit. The second group are classified as *output transducers*. Devices in this second group accept the electrical energy and convert it into a form of energy acceptable to the human mind, either through sight or hearing, or both.

Information is presented to the amplifier, radio, tape recorder, or television in many forms. Each type of device is designed to receive only some of these forms and to reject or ignore the other forms. Certain physical properties of each kind of device are utilized to convert the signal information into electrical energy. The items discussed in this section are typical of common input-transducers. Photographs of these devices are included as well as their symbols. These are the symbols used to represent the devices on the electronic schematic or wiring diagrams that are used by technicians to see how these units are wired.

Microphones are devices that convert sound waves into electrical energy. This conversion is accomplished when the sound waves exert a force on the microphone *element*. The element is designed either to produce electrical energy or to modify existing electrical energy in order to produce the desired electrical waves. When the waves are produced, they may be observed on an oscilloscope—a device used to display the height and shape of electrical waves. Waveforms discussed in this book are similar to ones observed on the oscilloscope.

Microphones are made in different ways to meet the different requirements of users. They are made in several shapes, and each contains one of several different kinds of elements. While the shape of the microphone case may be critical in order to pick up all the required sounds, it is the element that converts the sound into electrical waves. Types of elements commonly used in microphones are the carbon, crystal, or magnetic.

In order to better understand how these elements work, one must have an understanding of how an electrical circuit works. Since we are studying the movement of the electron in this book, let us first take a look at how the electron, in order to do some work, moves in a circuit. The *electron* is a negatively charged particle and will move when there is some outside force or pressure to make it move. When it does move, it will move from an area that is highly concentrated with other electrons to an area with fewer electrons. Areas with comparatively few electrons are considered to have a positive electrical charge. Therefore, electrons move from negative to positive, in a circuit. The electron will take the easiest path available in order to reach its goal. If there are multiple paths, electrons will move in all of them. The number of electrons moving in any path depends upon two factors. These are the amount of pressure or force available and how much electrical opposition there is in the path. Translating these statements into electrical terms, the electron movement is called the electrical *current*. The pressure is called the electromotive force or *voltage*. The opposition to electron movement is called *resistance*.

In any circuit, if current is moving through a resistance, a voltage will develop across the resistance. In a circuit containing at least two resistive units and only one path for electron movement, the voltage develops across each of the resistors in direct proportion to the amount of resistance in each of the resistors. If one resistor has a higher value of resistance than the second resistor, then the largest amount of voltage will appear to develop across the higher resistor value. By this same rule, the smaller resistor will have a smaller *voltage drop* develop across it. The exact amount of voltage developed across each resistance depends upon the amount of applied voltage and the relationship of all resistances in the circuit. The interaction of these factors will be covered in another chapter. First, let's look at various kinds of input transducers to see how they are made and how they work electrically.

The carbon microphone circuit. This microphone consists of an electrical circuit made up of a power source, a sensing unit, and the carbon granules in the element, as shown in Fig. 2-1. The voltage from the power source is divided proportionately across each of the two resistances in this circuit—the carbon granules and the sensing element. When there is no sound wave, the carbon granules lie loosely and their resistance is high in comparison to the resistance of the sensing element; thus, most of the voltage applied by the power source will be developed across the granules, and little voltage will be developed across the sensing element. Since the output of this microphone is taken from the sensing element, there will be no noticeable output. As the sound waves increase in strength, they cause the diaphragm to move and the carbon granules in the microphone element are compressed. Their resistance decreases, and so the voltage developed across them also decreases. An increase in voltage will then develop across the sensing element. In effect, there are two resistances in this circuit, with a voltage applied to the total circuit. There is only one complete circuit and only one path for electron movement. If one of the resistances decreases in value, then the voltage developed across each of the resistors will change in order to reflect this change in the resistor values. The voltage developed across both resistors is still directly related to the size of both resistances. A continual change in

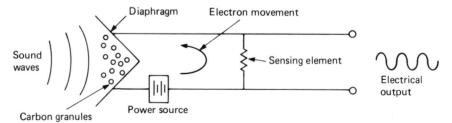

Figure 2-1. The carbon microphone circuit. Electron movement through the circuit causes a voltage variation to develop across the sensing element. These variations represent the sound waves and appear as electrical signals.

the resistance of the sensing element causes a continual change in voltage developed across the sensing element resistance.

Variations in the amount of sound waves reacting on the carbon granules will be translated into variations of the amount of voltage developed across the sensing element, and sent to the input of an amplifier, as shown in Fig. 2-2. These voltage variations are directly related to the intensity of the sound wave as it acts against the microphone element, translating the sound waves into an electrical signal voltage which is then processed further by the amplifier.

Sound Wave Intensity	Carbon Element Resistance	Voltage Developed Across Sensor
Low Medium High	High Medium Low	Low Medium High

Figure 2-2. If the power source voltage is kept at a constant value, voltage variations reflecting loudness will develop across the sensing element of the carbon microphone circuit.

The crystal microphone. The crystal microphone uses a different principle from that used by the carbon microphone to change sound waves into electrical waves. In the crystal microphone, a thin piece of crystalline material, usually quartz, is suspended between two supports. This material is capable of producing an electrical wave when its shape is changed. Electrons contained in the material move towards one side of the quartz as it is bent in one direction. Reversing the direction of bending causes the electrons to move toward the other side of the material.

Sound waves striking a diaphragm attached to a thin piece of quartz-crystal material will cause the crystal material to change its shape. The movement of electrons produced as the crystal's shape changes produces a voltage directly related to the amount of sound striking the diaphragm. A pictorial representation of this is shown in Fig. 2-3. As the sound wave increases in intensity, the voltage will increase in value. In this manner

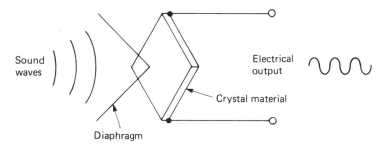

Figure 2-3. The crystal microphone circuit. Sound waves striking the diaphragm change the shape of the crystal. The changing shape generates a voltage at the output terminals.

variations in sound waves are converted into variations in electrical waves sent on to the input of the amplifier.

The magnetic microphone. The magnetic microphone utilizes yet another principle in order to transfer the sound waves into electrical energy. This principle is based on the relationship between a magnetic field and a piece of wire in the magnetic field. If the position of the wire in the field is changed, or the field itself is changed in strength or position, a voltage will be produced on the wire. The magnetic microphone consists of a permanent magnet formed into a U shape, a metal plate called a diaphragm which is suspended immediately in front of the ends of the magnet, and small coils of wire placed over the ends of the magnet, as shown in the Fig. 2-4. A magnetic field is produced by the magnet and surrounds the coils of wire. Sound waves striking the diaphragm cause it to move closer to the ends of the magnet, producing a change in the strength of the magnetic field. As the field changes strength, a voltage is produced by the moving field and develops on the wires in each coil. Constantly varying voltages, representing sound waves, are produced in this manner. Louder sounds produce a greater movement of the diaphragm and, in turn, cause a bigger voltage to be sent to the amplifier. Another name for the magnetic microphone is the dynamic microphone.

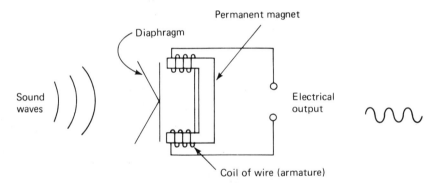

Figure 2-4. The magnetic, or dynamic, microphone produces a voltage representing the sound waves by means of changing the strength of a magnetic field in the microphone.

Schematic symbols. Construction of a car, boat, or building requires a set of plans to work from. This is also true of electronic equipment. A set of symbols of parts peculiar to the car, boat, building, or electronic device is used for representing its specific components. This saves many hours of drawing time for the planner, and symbolic representation has become universally used by persons engaged in repair as well as construction work. The electronics industry is no exception among industries using this sort of shorthand. Each electronic unit is constructed from an electronics blueprint called the *schematic*. The schematic uses *schematic symbols* to represent devices commonly found in each unit. The symbols are universal within the industry and have been standardized in the United States by the Electronic Industries Association. Often, a numerical value is used with the symbol to further identify the device. This is true when several similar devices with different values are used in one unit. The technician is thereby able to mentally picture the kind of device shown symbolically on the schematic. He will also be able to determine from looking at a device where it is represented on the schematic diagram.

Figure 2-5 shows several styles of microphone commonly in use. The schematic symbol is also shown. Often, only one schematic symbol is used to represent a general category of devices. The manufacturer's literature is utilized when it is necessary to determine the exact electrical and mechanical characteristics for a particular microphone.

2-3
The Phonograph
Cartridge

Two basic types of phonograph cartridge, or pickup, are currently in use. One of these uses the crystal principle and the other one uses the magnetic principle. When phonograph records are manufactured, each groove in the record has many wavy patterns in it. These patterns represent the music or whatever information is contained in the finished

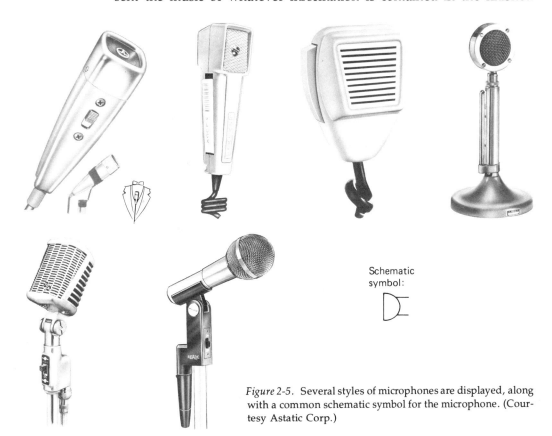

Schematic symbol:

Figure 2-5. Several styles of microphones are displayed, along with a common schematic symbol for the microphone. (Courtesy Astatic Corp.)

record. In order to reproduce the information from the record, a mechanical device is used to transfer the information to the transducer. This device is the phono *stylus,* or needle. The phono needle rides in the groove of the record. As the record revolves, the needle vibrates in response to changes in the waves in the record grooves. The information on the record is thus transferred to the phono cartridge where it is changed into an electrical signal by the element inside of the cartridge.

The crystal phono cartridge. The crystal phono cartridge is very similar in operation to the crystal microphone. The basic difference between these two devices is that the cartridge uses a phono needle, which changes its shape, while the microphone uses a diaphragm. The needle is attached to one end of a piece of quartz-crystal material. This is shown in Fig. 2-6. The movement of the needle in the groove of the record makes the crystal change shape. As the shape is changed, electrons move to one side or the other of the crystal. The needle movement in the record groove causes the crystal material to bend alternatively in two directions. The electrons move back and forth across the crystal material, producing a constantly varying voltage. This movement of the electrons produces voltage changes at the output terminals of the cartridge. The variations of voltage represent electrically the loudness and tone of the information recorded on the record.

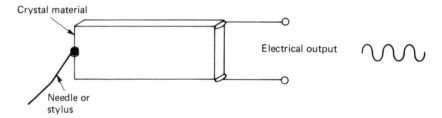

Figure 2-6. The crystal phonograph cartridge produces a voltage representing an electrical signal as the needle vibrates in the record groove. These vibrations cause the crystal to change shape and thus generate varying voltages.

The magnetic phono cartridge. The magnetic phono cartridge is similar in operation to the magnetic microphone. As with the microphone, a method of varying a magnetic field is used to produce voltage changes. The movement of the needle as it vibrates in the groove of the record sets up action within the cartridge, similar to that of a small electrical generator. The rate of movement of the needle is directly related to the amount of signal voltage generated. If the information on the record is loud, then the needle will generate a relatively large signal voltage. Soft music or vocal patterns will generate a lower electrical signal. A pictorial representation of the magnetic phonograph cartridge is shown in Fig. 2-7. The output connections from the cartridge are wired to the audio amplifier, in order to reproduce the information contained on the record.

The movement of a needle as it vibrates between the ends of a permanent magnet changes the magnetic field, and thereby produces a voltage in the armature windings, as indicated in Fig. 2-7. These variations are equal to the information recorded on the record and thus may be utilized to send information in the form of electrical voltage waveforms to the amplifier.

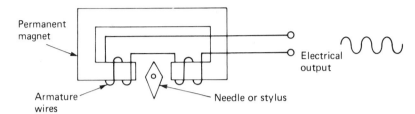

Figure 2-7. The movement of the needle changes the magnetic field in the cartridge, producing a voltage which represents the information on the record.

Phono cartridges are manufactured in several sizes and shapes. Pictured in Fig. 2-8 are several common cartridges and their common schematic symbol.

Figure 2-8. Phonograph cartridges are manufactured in various shapes. Several typical shapes are displayed here, along with their common schematic symbol. (Courtesy Astatic Corp.)

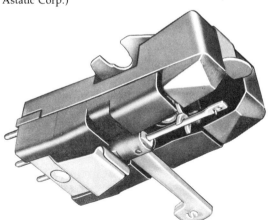

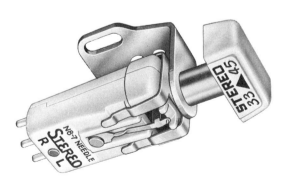

2-4
Tape Recorder
Heads

When a tape recording is produced, the sound information is transferred to the tape in the form of magnetic fields. Each field's shape represents a portion of the information recorded on the tape. A detailed discussion of this procedure is presented in another chapter. The *tape head* consists of a coil of wire wound about an iron core, which looks like, but is not, a magnet. This is shown in Fig. 2-9. As the tape moves past the tape head, the magnetic fields representing the information recorded on the tape have an effect on the head. Each magnetic field has its own force field around it. The force field, which is invisible to the eye, varies in size in proportion to the magnetic strength of the tape area. The force field is capable of generating a voltage in a wire even though the parts of the tape do not touch the wire. This generation is accomplished by an electrical principle called *electromagnetic induction.* In the case of the tape head, the induced voltage developed by the movement of the tape past the head is picked up by the coil of wire on the head and sent on to the amplifier. As in the case of the other transducers, variations in loudness and tone are reflected in the strength and shape of the electrical voltage waveforms sent to the amplifier. This voltage is usually referred to as the signal voltage, or as just the signal.

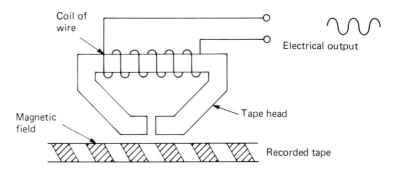

Figure 2-9. A varying magnetic field moving past the tape head induces a voltage on the coil of wire and at the output terminals of the tape head. The voltage is the electrical form of the information on the tape.

2-5
Antennas

Radio and television broadcasting stations emit electromagnetic waves from the transmitting tower or antenna of the station. These waves are sent out into the air in order to be picked up, or received, by the radio or television receiver. While receiving antennas vary in size and shape, they all work on the same principle. This is the principle of electromagnetic induction. As electromagnetic waves cross a piece of wire or metal, a voltage is induced in, or placed upon, the wire or metal. If the number of the waves is large, the induced voltage will also be large. Receiving antennas are designed by electrical engineers for the specific purpose intended—either to receive one particular type of transmitted wave or to receive several types of waves. Antennas built into radios and television sets usually are designed to receive several stations. Typical antennas used for radio and television reception, and their common schematic symbol, are illustrated in Fig. 2-10. After the electrical signal is received by the antenna and the appropriate signal voltage has been developed, this information is sent on to the radio or TV set for further processing.

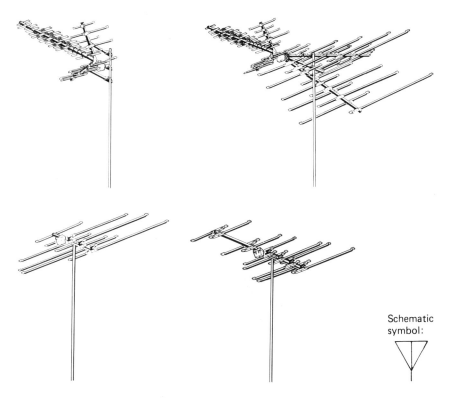

Schematic symbol:

Figure 2-10. Three typical receiving antennas are shown, along with a common schematic symbol for an antenna. (Courtesy Winegard Co.)

2-6
Television Cameras

The television camera converts a visual image into an electrical signal. Inside the camera is a special kind of vacuum tube which contains a light-sensitive plate. An image is projected onto this plate through the lens of the camera. The image changes the electrochemical composition of the plate, preparing it for another process. At the same time as the image is projected onto the light-sensitive plate, a beam of electrons moves across the plate, scanning it, so to speak. The electron beam is designed to move at a specific rate of speed, examining the image horizontally, one line at a time. After the beam completes one scan, electronic circuitry in the camera arranges for it to drop down to another line and repeat the process. As the beam strikes the light-sensitive plate, a change occurs in the amount of beam current flowing in the tube. This beam current is produced inside the tube and acts as a sensing element in the tube. Changes in the amount of beam current cause a change in the amount of voltage developed across a sensing resistor in the camera, thus converting the image into an electronic signal voltage for further processing. There is no standard schematic symbol representing a television camera.

2-7
Electrical Waves

Electrical waves are produced by a generating device. The quantity of waves the generator produces in one second is called the *frequency* of the generating device. These devices are divided into two basic categories. At slow speeds, or low frequencies, a mechanical generator is often used to produce electrical waves. Such generators may also be used to produce great quantities of electrical power.

As a wire, or loop of wire, is rotated in a magnetic field, a voltage is induced on the wire through the process of electromagnetic induction. Increasing the number of windings on the wire loop increases the voltage generated. Power companies use this principle to generate electrical power for use in homes and industries. Due to the large size of this type of generator, the speed is kept low. Otherwise, the generator could self-destruct due to the strong centrifugal forces created as it rotated at a high speed.

High frequency generators normally use electronic vacuum tubes or transistors to produce electrical waves. By varying the components in the electronic generator, the frequency and size of the waves produced may be modified. Three types of waves are in common use in electronics. These are the sine wave, the square wave and the sawtooth wave. The schematic symbol for a generator is a circle. The circle, representing the generator, contains the shape of the type of wave it generates, as shown in Fig. 2-11.

Sine wave   Square wave   Sawtooth wave

Figure 2-11. Three basic electrical waveforms commonly used in electronic work. These are the sine wave, the square wave, and the sawtooth wave. The circles represent generators.

2-8
Output
Transducers

After the signal information has been accepted or received by the unit, it is processed. The manner in which it is processed is detailed in later chapters. The electrical signal in its final form must be converted into either a visual image, or into sound, in order to be received by human sense perception. This conversion of electrical signals into other forms of signals is accomplished by another kind of transducer, classified as an output transducer. An output transducer works in a manner opposite to that of an input transducer. Output transducers commonly found in home entertainment equipment are the loudspeaker and the cathode-ray tube.

2-9
The Loudspeaker

The loudspeaker works on the principle that two magnets when close to each other tend to interact. We have heard that "opposites attract, and likes repel." This is true of magnets, whether they are permanent magnets, electrical magnets, or both. Controlling the amount of movement of the paper cone in a loudspeaker is the method used in producing sound waves from electrical waves. A coil of wire, called a voice coil, is placed around a permanent magnet. The permanent magnet is attached to the frame which is around the loudspeaker. The voice coil is attached to a movable paper cone whose outer edges are also attached to the loudspeaker frame, as shown in Fig. 2-12. Wires from the voice coil are connected to the output stage of the radio, tape recorder, or television set. Electrical signals from the set change the magnetic field in the voice coil, causing it to move either toward or away from the permanent magnet. This movement is so powerful that sound waves are produced. Variations in the strength and frequency of the electrical signals produce tones we are able to hear. The loudspeaker works in a manner similar to the magnetic microphone, only in reverse. Pictures of typical loudspeakers are shown, along with a common schematic symbol, in Fig. 2-13.

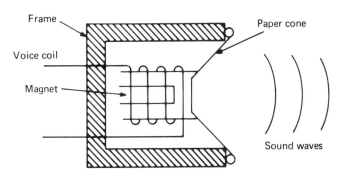

Figure 2-12. Interaction between two magnetic fields causes the paper cone in the loudspeaker to move. The cone moves at a rate which will cause a production of sound waves, thus reproducing the information being amplified.

2-10
The Cathode-Ray
Tube (CRT)

The cathode-ray tube works on the principle that, when a beam of electrons strikes a phosphorescent surface, that action causes the surface to glow. The CRT, as it is often called, has a front plate which is coated on

Schematic
symbol:

Figure 2-13. Several sizes and shapes of loudspeakers are displayed, along with their common schematic symbol. (Courtesy Quam-Nichols Co.)

the inside with a phosphorescent material. The chemical makeup of this material determines the color of the glow. CRT's used in electronic testing equipment are usually coated with materials that will produce a green color. Black-and-white TV sets have tubes which show white when struck by the electron beam. Color TV sets produce three basic colors —red, blue, and green. All CRT's accomplish this by the generation of a beam of electrons at the back of the CRT, in a device called an electron gun. The gun produces free electrons, and after these electrons are shaped into a fine beam, they are directed toward the face of the CRT, as shown in Fig. 2-14. Other circuitry associated with the CRT, and the rest of the unit, causes the beam to sweep from side to side across the face of the CRT and to move up and down the face of the CRT as well. Another

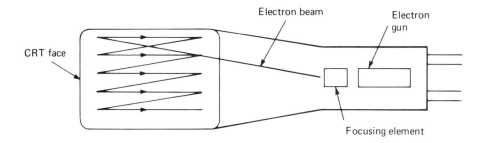

Figure 2-14. Electrons generated in the gun section of the CRT are directed against the face of the tube. As the electron beam moves across the tube, it *paints* a picture, one at a time.

circuit controls the intensity of the electron beam. As the beam moves across the face of the CRT, one line at a time, it literally paints a picture. Movement and intensity are synchronized with the TV camera and transmitter thus permitting the viewer to receive an image similar to the one sent out from the broadcast station. The ability of the human eye to retain an afterimage produces an illusion of a complete picture to the viewer.

The simplest CRT is a one-gun type. One electron beam is generated and one resulting image is displayed on the face of the CRT. This device is used in black-and-white TV sets and in common electronic-testing equipment, such as the oscilloscope. It is possible, however, to construct a more complex CRT having two or three guns. Dual-gun CRT's are often used in oscilloscopes when it is necessary to observe two different electrical signals at the same time. In some cases, an oscilloscope uses an electronic switching circuit rather than a dual-gun CRT to display two signals at the same time. These oscilloscopes are called dual trace units. They are found in laboratories and are currently being utilized by radio, television, and sound system repair technicians due to the increased complexity of modern electronic home entertainment devices.

A three-gun CRT is used in most color TV sets. The coating on the inside of the face of a color CRT consists of sets of three dots. One dot in each set glows red, one glows green, and the third dot glows blue. Three electron guns are required in order to produce a color picture. Each gun has its own control circuitry for intensity and position of the beam. The three beams move as one in order to land on the same set of phosphor dots at the same time. Varying the intensity of the beams produces the variations of color seen on the screen. How this is accomplished is covered fully in a later chapter. The schematic symbols for one-, two-, and three-gun CRT's are shown in Fig. 2-15.

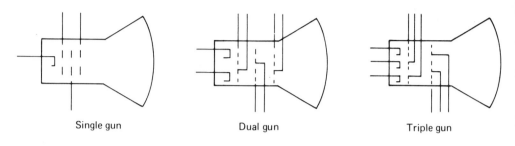

Single gun   Dual gun   Triple gun

Figure 2-15. Schematic symbols for one-, two-, and three-gun CRTs are shown. Use of one tube outline for all of the elements indicates that all are contained in a single unit.

SUMMARY

Home-entertainment electronic devices process electrical signals. Some mechanism is required in order to change sound or radio waves, or light, into an electrical signal, or to change electrical signals into sound waves or pictures. These devices are called transducers. Their basic function is to change one form of energy into another form of energy. For the purposes of this book these devices are limited to input transducers and output transducers.

Typical transducers used as input devices for home entertainment are the microphone, the phonograph cartridge, the tape recorder head, and the television camera. Two devices commonly utilized as output transducers are the loudspeaker and the cathode-ray tube. Each of these is produced in a variety of physical sizes and shapes.

Persons employed in electronics use a graphic shorthand to represent these devices on schematic diagrams. They have to be able to identify each basic device and its schematic symbol in order to be effective in their work. Learning how each device functions and what it looks like, along with learning to identify common schematic symbols, are necessary steps in the development of a qualified repair technician.

QUESTIONS

1. What is meant by the term *transducer*?

2. What kind of work is performed by an input transducer?

3. What kind of work is performed by an output transducer?

4. What electrical principle is used in each of the following transducers?
   (a) carbon microphone
   (b) loudspeaker
   (c) phonograph crystal

(d) cathode ray tube
(e) dynamic microphone
(f) tape recorder head
(g) antenna

5. What are the three basic electrical waveforms commonly used in electronics?

Section Two

Block
Diagrams

Block Diagrams: Audio Devices

3

Technicians involved in the repair of home-entertainment electronic equipment need to know how the particular device works. Assuming that the manufacturer built a working device, but that some part of the device has failed, the repair technician must be able to localize the trouble to a specific area. Once this is accomplished, he then is able to make tests and measurements in the device in order to locate a nonfunctioning part. The step from looking at or listening to the device to locating a specific malfunctioning part is a large one. It is normally easier to add some *in between* step. Most often this is accomplished when the technician identifies a specific block, or area, in the unit in which the trouble may be found. Actually, the entire unit may be thought of as consisting of *blocks*. Almost every AM radio has similar sections or blocks. Any black-and-white television set has blocks that are commonly found in most TV's regardless of which company produced the set. These similarities enable the technician to analyze the symptoms seen or heard, to determine which areas or blocks may be nonfunctioning, and then, by reference to the manufacturer's service literature, and use of electronic testing-equipment, to make a successful repair.

The importance of learning how each type of set is made up of similar blocks cannot be over-stressed. Some of these blocks are found in many kinds of sets. An example of this is the power-supply block. It performs the same kind of job regardless of the type of set in which it is used. There are differences in electronic circuitry to be found in the design of different power supplies, but one can assume that the function of a power-supply block is to meet the operating power requirements for the set. Electronic circuitry will be covered in detail in later chapters. This chapter covers the audio-amplifier block diagram, tape recording processes, and tape players. Once the block diagram of an audio amplifier is understood, the principles involved may be applied to other kinds of devices that use audio outputs, such as radios and television sets.

3-2
The Audio
Amplifier

When discussing electronic theory, it is wise to have a common language. Special words or phrases in everyday use by people working in electronics may at first seem difficult to understand to persons learning about this subject. As these words or phrases are introduced in this book, the author's definitions will be given. In this way, the reader will be able to better understand what the author has in mind as he presents the material covered.

In a discussion of the audio amplifier, one needs to know what is meant by both *audio* and *amplifier*. To amplify means to enlarge. As we know, this book studies electrical signals. An amplifier enlarges electrical signals. The amount of amplification will vary from two to three times the original signal, in a simple amplifier, to several-thousand times in a complex amplifier. The amount of amplification is based upon the requirements or usage of the amplifier and is determined by the engineer designing it.

The electrical signal has certain characteristics that should be mentioned at this time. One of these is the *amplitude* of the signal. This term refers to the intensity or strength of the signal. It is usually measured in volts. Another characteristic is the rate of repetition. This refers to the number of times, in a given time period, that the wave is repeated. These characteristics are illustrated in Fig. 3-1. The standard time reference for electronic signals is one second. One complete wave is called a *cycle*. The *frequency* of the waves in the time frame of one second used to be called *cycles per second*. This term has recently been changed, to conform to international usage, to *hertz*. One hertz equals one cycle per second. The term is named in honor of the German scientist who demonstrated this.

Figure 3-1. The basic characteristics of an electrical signal include amplitude and frequency.

An acceptable form of usage is to state, for instance, "The frequency of that signal is 120 hertz." In earlier years, one would have said, "The frequency of that signal is 120 cycles per second."

Electrical waves are grouped according to their frequency and use. The chart shown in Fig. 3-2 gives a breakdown of the general groups. The audio frequency range is at the low end of this chart. These frequencies are those which are said to be heard by the human ear. Generally speaking, the *audio frequency* range is from 10 hertz to 20,000 hertz. Frequencies above these are categorized under the broad term of *radio frequencies*. A breakdown of the radio frequency range gives us low, middle, high, very high, and ultrahigh frequencies. These last two are used in TV sets and commonly referred to as *VHF* and *UHF*. There are useable frequencies that are higher than these, but they will not be covered at this time.

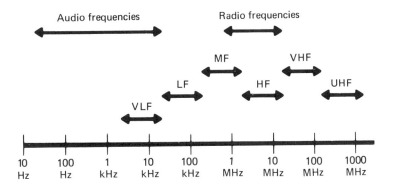

Figure 3-2. An analysis of the frequency spectrum as used by home-entertainment devices. Low frequencies, starting at 10 hertz and increasing to 20,000 hertz, are audio frequencies. Higher frequencies are divided into radio and television bands.

Going back to the terms *audio* and *amplifier* and the phrase *audio amplifier*, we are now able to define an audio amplifier as an electronic device that is capable of enlarging an electronic signal in the audio frequency range. The amount of amplification will depend on the design of the amplifier.

Amplifiers are not limited to audio frequencies. They can be designed to work for all other frequencies of electronic waves. An audio amplifier, however, is designed to work with frequencies that are in the audio range. An audio amplifier may be considered as a unit, as shown in Fig. 3-3(a). Of major consideration in this unit are the input and output transducers. So far, we haven't considered what is inside the unit. A repair technician would have a very difficult time attempting to repair this amplifier unless he had further information. One of the first steps in making a repair would be to break the unit down into its blocks, as shown

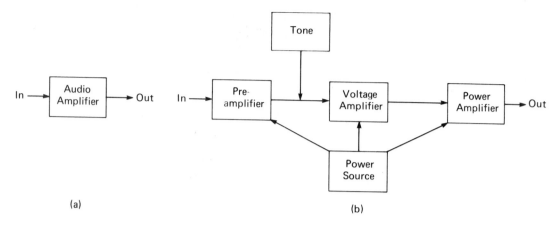

Figure 3-3. (a) The audio amplifier unit. (b) The same unit broken down into functional blocks.

in Fig. 3-3(b). From this, we derive the term *block diagram*. Each unit consists of its functional blocks. Later on, each block will be broken down into (1) how it processes the electrical signal and (2) what electronic circuitry is in it. For now, learn the basic function of each block. Input and output devices were covered in Chap. 2, and there is no need to repeat that information. Following is a discussion of the blocks in the audio amplifier.

The preamplifier.   Electrical signals from input devices such as microphones or tape-recorder heads are very low in signal amplitude. In order to increase the amplitude of this signal, a preamplifier is used. A preamplifier enlarges the signal voltage to some predetermined level, or amount. This is done in order to have the electrical signal conform in amplitude to other input signal levels processed by the following stages of the amplifier. Under normal operating conditions, the signal level at the input of a preamplifier stage is very low and requires amplification in addition to that provided by the set in order to be heard in the speaker. In a situation where an amplifier may have several input capabilities, such as tape recorder, phono cartridge, AM and FM tuners, and a microphone, the amplifier unit may use a preamplifier block with those input devices that have lower output signals than others, in order to provide equal output signal levels at the speaker.

Tone controls.   Circuitry in this block is used to modify the electrical signal in order to emphasize the high frequencies (treble) or to emphasize the low frequencies (bass). This block's controls may also be used to de-emphasize either high or low frequencies. Turning tone-control knobs can result in the production of mostly middle and high frequency tones or mostly middle and low frequency tones, depending on how the controls are adjusted.

The voltage amplifier. The signal produced by the input transducers, and processed by the preamplifier and tone controls, is in the form of an electrical voltage. This voltage is very small, on the order of 0.0005 to 0.001 volts. It must be further amplified in order to be able to perform the necessary work. This is accomplished by processing it through a voltage amplifier. The electrical output of the voltage amplifier may be on the order of 1.0 to 1.5 volts. This amounts to an increase of up to 2000 times the input signal level. (The amount of amplification achieved by a specific amplifier will depend upon its design.) In order to do this much work, the voltage amplifier may contain many transistors or tubes. The contents of each block in a block diagram may be very complex. Later chapters of this book discuss what makes up the amplifier, and the other kinds of blocks. At this time, we are looking only at how each block functions within the unit. The basic function of the voltage-amplifier block is to enlarge the electrical-signal amplitude sufficiently to operate the next block of the amplifier.

Power amplifier. Power amplifiers are required in order to produce the necessary power requirements (voltage and current) to operate the output transducers. Power-amplifier blocks are designed to increase the electrical power output of the amplifier. A typical output-transducer, such as a loudspeaker, requires a low voltage and a high electrical current in order to create the magnetic field in the speaker. Its power amplifier is designed to meet these needs. Large amplifier systems, such as are used by musicians and for public address systems, need great quantities of electrical power in order to reproduce the original sounds. In general, the output of the power-amplifier stage must be matched electrically to the output transducers. If this is not done, a mismatch occurs and much of the power produced in the amplifier is wasted.

Power source. Electronic devices need electrical energy in order to function. This energy is used to do useful work in the system or unit. The functional unit that accepts the energy from an outside source and converts this energy into the proper operating voltages for the unit is called the power-source block. The block may obtain the required voltage and current from a battery, as in a portable unit. More complex units obtain power, generated by a local electric company, by plugging the unit's power cord into an electric wall-outlet. The power source block used in a unit which plugs into a wall outlet is more complex. In simple terms, the power source accepts electric power and converts this power to establish the proper operating voltages and currents for the components in the set.

The unit under study so far has had only one channel. It is classified as a *monaural* unit (mono means *one,* aural means *sound*). Today, we have *stereo* units on the market. These are either two-channel or four-channel devices. Stereo units, as illustrated in Fig. 3-4, are made up of two audio amplifiers working as one unit, with two inputs and two outputs, all

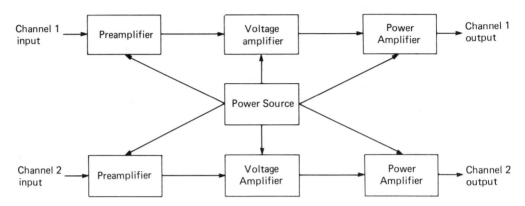

Figure 3-4. The stereo amplifier, in block form. Each channel has the same kind of functional blocks. There is usually one common power-source block for the unit.

operating at the same time. Typical recording procedures record sounds into a *right* channel and into a *left* channel. The right-channel information and the left-channel information are recorded on the record, or tape, in a manner permitting both channels of information to be played back at the same time. Dual-element phono cartridges or tape-recorder heads are used in this system in order to play back both channels simultaneously. The output of each of the elements in the input transducer is connected to a separate input on the stereo amplifier. Right-channel information is amplified in the right-channel amplifier blocks. Left-channel information is amplified in the left-channel amplifier blocks. Each channel's output is connected to a separate speaker. Normally, the speakers are placed in different corners of a room, as shown in Fig. 3-5(a). The listener sits between the speakers, but toward the wall opposite to where the speakers are placed. Listening in this manner gives the effect of sitting in an auditorium and listening to a *live* concert.

Quad (short for quadraphonic), or four-channel, sound uses the principles employed for two-channel stereo and adds some refinements. A perceptive listener notes a slight echo effect when hearing a live concert. This effect is re-created with a quad stereo system by adding two additional amplifier channels. Each amplifier unit has its own speaker system. A typical installation for quad sound would place one speaker in each corner of the listening-room. The listener sits near the center of the room, or possibly towards the rear, as shown in Fig. 3-5(b). Most of the sounds come from the two front-speakers. There is a reduced level of sound

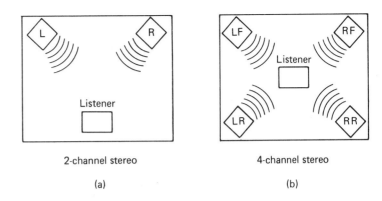

Figure 3-5. Two-channel stereo requires two speaker-units, one for each channel. Four-channel stereo adds two additional speakers, which are placed at the rear of the room. This gives the listener the impression that he is in the center of the room in which the music is being played.

from the rear speakers. These simulate the echo effect found in an auditorium, in which sounds are reflected back from the walls at the rear of the auditorium.

The units for two-channel and four-channel systems can be purchased separately, or they are also available with all channel amplifiers and power sources enclosed in one common package. Some units use one common power source for all of the channels.

3-4
Tape Recorders
and Players

Tape units have become big business in the past few years. The biggest sales seem now to be in the area of playback units for auto and home use. Understanding the process of putting music or words on tape is part of understanding how the entire device works. Recording tape is basically a strip of plastic with some chemicals glued to one side of it. See Fig. 3-6. These chemicals are a special group of oxides and may be easily magnetized. The combination of plastic and oxide must meet certain basic requirements. It must be flexible so that it will fit snugly against the tape recorder heads as it passes them; it cannot stretch, because then the oxide coating could fall off of the plastic; it must be smooth in order to move across the heads without damaging them with scratches; and it has to be thin in order to obtain adequate recording time (otherwise, it would occupy too much space). How it works is fairly simple. When the unrecorded tape moves past the recording head, the oxide particles are magnetized in a pattern that represents electronically whatever sound-information one wishes to record. Each particle of oxide—and there are literally billions of them—receives some magnetism. The magnetism causes these particles to line up in a specific pattern representing words, music, or both. The magnetic force which causes the oxide particles to line

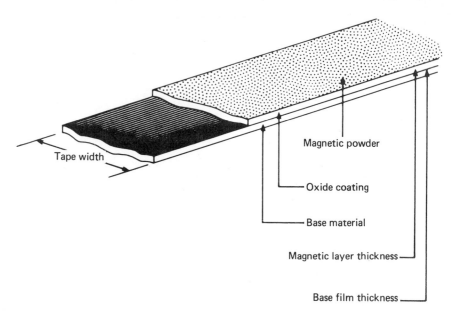

Figure 3-6. Recording tape is made up of layers of material consisting of a base material, an adhesive, and an oxide coating. The oxide coating consists of magnetic materials which contain the sound information that is recorded.

up comes from the tape head. See Fig. 3-7. Entertainment devices use frequencies in the audio range, but units used for industrial purposes operate on the same principle, using higher frequencies.

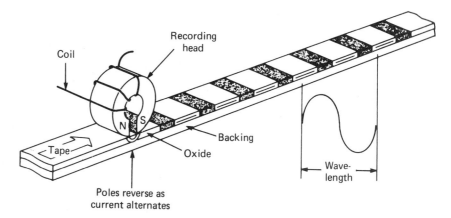

Figure 3-7. How a tape is recorded. A magnetic field is generated in the coil of the recording head. This field arranges the oxide particles on the tape to represent the information being recorded.

Track systems. Recording tape used with home-entertainment record-playback units may contain from one to eight channels of information. The number of tracks used is directly related to the length of the playback time and to the quality, or fidelity, of the recording. These two factors, time and quality, are evaluated when the tape is manufactured and help to determine the kind of playback system purchased by the consumer.

Generally speaking, the wider the track on which the information is recorded, the better the quality of reproduction. This is true because as the recording-track width is reduced, the higher frequency notes cannot be recorded on the tape. Choosing track width becomes a compromise between the number of tracks to be used and the desired quality of the recording. Many of us, as we age, lose the ability to hear the higher audio frequencies. Therefore, a lack of ability to record the high frequency notes becomes less critical.

Tapes used in the home-entertainment field today have from one to eight tracks of information on them. Most of the tape record-playback systems in current use use either two, four, or eight tracks. Methods of recording have had to be worked out in order to handle the additional tracks of information on the tape. One thing that was done was to reduce the width of the tape head and thus be able to reduce the width of the recorded track. In a half-track system, head width is reduced to slightly less than half of the width of the tape. The head is then mounted in such a manner that it is offset from the center of the tape as it passes by the head. This combination of narrower recording width and offsetting the relationship between the head and the tape produces a recording pattern as shown in the Fig. 3-8(a). After half the tape has been recorded, it is

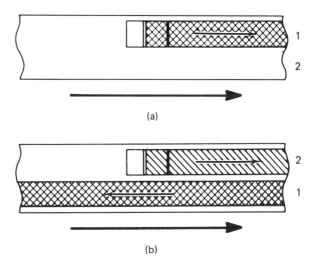

(a)

(b)

Figure 3-8. A two-track record-playback system. Each track is offset and utilizes half of the tape width. When one side is recorded, the tape is turned over on the machine. Recording occurs in one direction only.

removed from the recorder and reversed on the machine. Then it is run past the head again. During the second run-by, the information is recorded on the other half of the tape, as shown in Fig. 3-8(b). A system of this type permits two channels of information to be recorded on a single tape.

The introduction of stereo recordings has generated the requirement of using two separate tracks at the same time in order to record the stereo information. Each track in the pair contains one channel of information. Two recording heads are used in this system. The heads may be contained in the same casing, but electrically they are separate. Each head records its own set of information on the tape. Here, as with a two-track system, the heads are offset with respect to the position of the tape. The system commonly used for recording two-channel stereo is shown in Fig. 3-9. Each pair of tracks is recorded on as the tape moves in the first direction. After the tape has completed its travels, it is removed from the recorder and reversed on the machine. Then the other pair of tracks are used. Each track is separated from the adjacent track by a small area with no recording on it.

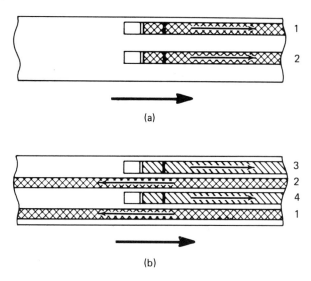

Figure 3-9. A four-track system. Tracks 1 and 2 are recorded as the tape moves. Turning the tape over on the machine permits recording on tracks 3 and 4.

Another track system that has gained wide acceptance in the past several years is the eight-track system. The eight-track system uses a nonreversible tape cartridge. Because of this, all tracks are recorded, as the tape moves, in one direction. The individual tracks are narrower than those used with other tape-recording systems. Pairs of tracks are recorded with stereo information. When using an eight-track system, the four pairs of tracks are used to record four programs. In this system the tape heads are moved to different positions instead of changing the positions of the tape. An indexing mechanism is used in the tape player in order to shift the head position at the end of each program. The tape has a metallic contact-strip on it at the end of the program track. When this contact strip completes its circuit in the player, the index mechanism shifts the head to a new position. Figure 3-10 shows how each pair of tracks is positioned on the eight-track tape. Using an eight-track system permits four times the playing time for one tape over a conventional two-track system.

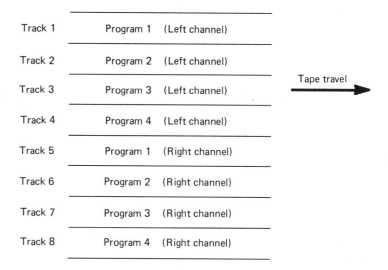

Track 1	Program 1	(Left channel)
Track 2	Program 2	(Left channel)
Track 3	Program 3	(Left channel)
Track 4	Program 4	(Left channel)
Track 5	Program 1	(Right channel)
Track 6	Program 2	(Right channel)
Track 7	Program 3	(Right channel)
Track 8	Program 4	(Right channel)

Tape travel →

Figure 3-10. The eight-track system. The record-playback head changes position after each pair of tracks is used. Direction of tape travel does not change in this system.

Other factors besides the number of tracks that influence the amount of material that may be recorded on the tape are the length of the tape and the speed at which it is played. Open-reel tapes are manufactured in standard lengths of 600-, 900-, 1,200-, 1,800-, 2,400-, and 3,600-foot lengths. The thickness of the plastic backing material is a prime factor in determining the length of the tape. The other tape containers, cassettes, as well as cartridges, are manufactured to play for specific lengths of time. Both the cassette and the cartridge playing machines are designed to operate at a single speed. Open-reel machines often have a multiple-speed mechanism. Cassette tapes are available in 30-, 45-, 60-, 90-, and 120-minute lengths. Cartridge tapes may be purchased with 27-, 45-, and 90-minute playing times.

Reel-to-reel systems. The first tape recorders made used open-reels for holding the tape. These reels were, and still are, made in a manner similar to those used for movie film. One reel is considered the supply reel. It supplies the nonrecorded tape to the recording mechanism. The other reel is empty when placed on the machine, and it is used to take up and hold the recorded tape. After the tape is recorded, the reels are rewound, thus putting the recorded tape back on the supply reel, ready for playback. If a four-track stereo system is used, then the placement of the reels is reversed on the machine and the tape is run through the recorder a second time, recording on the second set of tracks. There are many tape recorders on the market today using this open-reel concept. Tape speeds for reel-to-reel systems are 1-7/8, 3-3/4, 7-1/2, and in some cases, 15 inches per second.

Cassette systems. The cassette tape is a smaller version of the open-reel tape system excepting that the cassette is placed on the tape recorder as one unit. The tape does not need to be threaded past the recording heads. See Fig. 3-11. In cassette systems, the recording heads are move-able and move against the cassette tape when the machine is turned on. The front of the cassette has three openings. The left opening is where the erase head fits, the record-playback head fits into the middle opening, and the drive mechanism fits into the right-hand opening. Tape speed in a cassette system is 1-7/8 inches per second. The tape width in a cassette is about half that of the standard 1/4-inch-wide tape used on an open-reel system. The rear of the case of the cassette contains two square, plastic tabs, one at each end of the case. These tabs are part of a protection system for the information recorded on the tape. If the tabs are broken and removed, then the tape cannot be erased or rerecorded. A safety feature in the engaging mechanism locks out the *record* button so that it cannot be pressed down if the left-hand tab is broken off or missing.

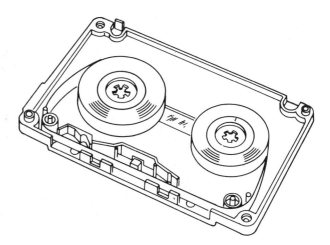

Figure 3-11. The cassette is a complete unit. It has only to be placed on the recorder in order to be used. There is no need to arrange the tape next to the heads. It resembles the open-reel system.

The cartridge. The eight-track cartridge contains an endless loop (that is, a *closed* loop) of tape, as shown in Fig. 3-12. The tape is a standard 1/4-inch width and runs at a speed of 3-3/4 inches per second. The cartridge also has three openings on its front to accommodate the erase head, the record-playback head, and the drive mechanism. In this system, a pressure roller against which the capstan, or drive shaft, is placed is contained in the cartridge. The capstan and pressure roller combine to pull the tape from the supply reel. Since the tape is endless, the supply reel also acts as a takeup reel. The tape is pulled from the center of the reel and is taken up on the outside of the reel. In this system the tape slips continuously against itself requiring that it be properly lubricated in order to minimize friction and breakage. If the lubricant becomes used up, then the tape will jam and probably break. On the other hand, too much lubricant will impair the ability of the head to record or play back the required information.

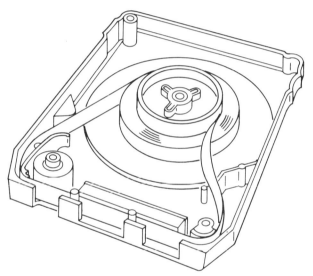

Figure 3-12. The 8-track cartridge has one supply-takeup reel. Tape lubrication becomes an important factor in its use.

Speed and frequency response. The selection of a tape speed often becomes a compromise. Two factors have to be considered. One is the amount of time available for recording. The second is the frequency response. These two factors are interrelated. Generally speaking, the higher the tape speed, the better the frequency response, or quality, of recording. Broadcast stations often use tape speeds of 15 inches per second when recording music programs. Such speeds will provide the best reproductions of the 10-hertz to 20,000-hertz audio frequencies. However, a large reel of tape must be used in order to record for even a 15-minute period. Cartridge players are not often required to reproduce the full range of audio frequencies. These units are often limited to frequencies of 8,000–10,000 hertz. This lower frequency requirement allows a longer recording time for a given length of tape. A compromise in fidelity is made in order to obtain the longer playing time.

Cassette manufacturers are also confronted with this problem. The cassette speed is 1-7/8 inches per second, which does not normally allow for reproduction of the higher audio-frequencies. Efforts to overcome these limitations have led to the production of magnetic oxides, for use on tapes, which offer much better frequency response. There are various trade names for this quality of tape, too many to mention here. Each tape manufacturer, however, advertises these products, and so they are not difficult to find.

The tape-recorder block diagram. Three types of tape recorder systems have been covered so far. While these systems have identifiable differences, they also have some basic similarities. Figure 3-13 is a diagram of the parts basic to all tape recorders. This diagram is not a true block-diagram, but it shows in graphic form the various components required in a tape recorder-playback unit and how they are physically placed on the mechanism. All tape recorders have a similar mechanical system consisting of a supply reel, a takeup reel, and a capstan. The capstan is usually connected to the motor and revolves at a constant speed. The particular speed used depends on the required recording speed. The pinch roller keeps the tape tight against the capstan, helping to keep the tape speed constant. When the machine is in a rewind or fast-forward position, the capstan is not used; instead, a drive mechanism connected to either the supply reel or the takeup reel is used in these positions.

The electronic portion of the tape recorder consists of an audio amplifier, a bias oscillator, and a power source. The audio amplifier consists of four blocks. These are the preamplifier, the voltage amplifier, the power source, and the power amplifier.

This preamplifier performs in the same manner as that found in other audio amplifiers. It receives an electronic signal from the tape head and amplifies it in order to produce a larger signal. The signal received from the record-playback head is very low—usually around 0.0005 volts. The preamplifier stage has to do a large amount of amplifying in order to get this signal voltage large enough to be processed by the rest of the amplifier.

The voltage amplifier increases the amplitude of the signal from the head still further in order to provide sufficient signal levels to operate the power amplifier.

The power amplifier does not increase the amplitude of the signal further. Instead, it works as a matching system and adds the proper amounts of power in order to operate the speaker.

The power source block is used to develop the proper operating voltages and currents for the motor and electronic portions of the set.

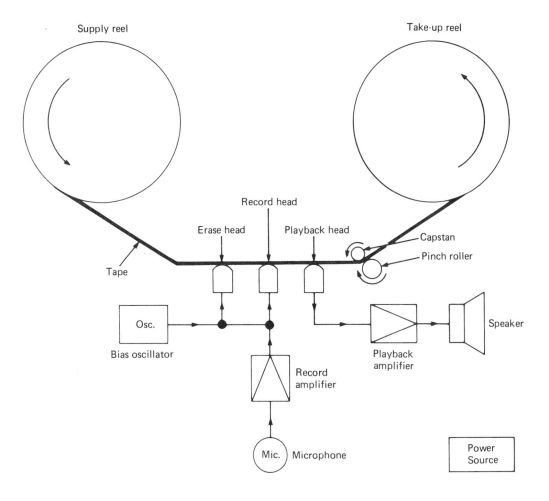

Figure 3-13. Component placement for tape record-playback units. All tape units use a form of this system. Often a combination record-playback head will be used instead of separate heads.

These blocks are contained in the playback amplifier block illustrated in Fig. 3-13.

One additional block is shown. This block operates only when the unit is in a recording position. It is turned off during playback periods. The block that has been added is the *bias oscillator*. An oscillator is an electronic device that is able to produce an electronic signal of a specific frequency. The exact frequency is determined by the components used to build the oscillator. This will be explained in detail in a later chapter. All that it is necessary to know at this time is that an oscillator is a device that will produce an electronic signal. The term *bias* is used in electronics to identify a voltage of some kind that is used to help operate the system. Often the bias provides certain voltages that make the system operate more smoothly. Tape-recorder bias performs two jobs. One is to erase or disorganize in an orderly fashion any information that might be found on a tape. After passing by the erase head, the tape is *clean* and ready for recording. Note that the erase head is the first one in line on the recorder. The second purpose of the bias oscillator is to provide a frequency to the record head in order to make a distortion free recording. The bias oscillator signal is mixed with the information to be recorded on the tape. In effect there are two electronic signals on the recorded tape. The bias oscillator and the music or voice information form these two signals.

Tape players. The tape player is very similar to the tape recorder. Instead of an erase head and a record head, the tape player uses a playback head and does not have the ability to record on the tape. All of the electronics required for recording purposes are left out of the unit when it is built.

Figure 3-14 shows a block diagram for a tape player. There is little difference between this and the basic audio-amplifier block diagram. Some tape players are made to be connected into home sound-systems. Such units do not require a voltage amplifier, power amplifier, or loudspeaker. The output of the tape player is connected through electrical cables to the input of the sound system. Units used for this purpose are called *tape decks.* They contain all of the mechanical and drive systems, but have only a preamplifier for the electronic section.

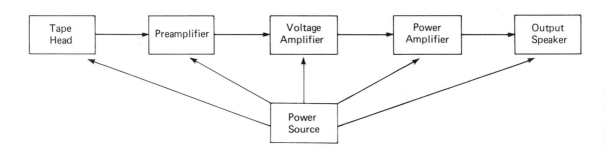

Figure 3-14. A typical tape-playback block diagram. Electronic signals are picked up from the play head and amplified by the system. The output is connected to an audio amplifier in order to operate a loudspeaker.

Because of the relatively low cost of producing a combination record-playback unit, most units found in the home or as portable units are combination record-playback units. By means of a switching circuit, the same head is used for either record or playback purposes. The electronics in the machine and the connection to the head are switched by means of an electromechanical device from the recording mode to the playback mode.

SUMMARY   Repair technicians need to know how electronic systems function if they are to be successful in their work. Knowledge of the block diagram of a set and how each signal is processed through the blocks is a basic requirement for the technician. Audio signals are enlarged, or amplified, in order to operate loudspeakers. The basic blocks for an audio amplifier are a preamplifier, a voltage amplifier, a power amplifier, tone controls, and a power source. Two- and four-channel systems use multiple amplifiers in order to reproduce the required information.

Tape recording is one process for storing and reproducing information. The tape consists of a plastic backing material coated with a magnetic oxide. The oxide is magnetized by the action of the tape head. Basic systems in use are the reel-to-reel, cassette, and cartridge. Each has its own advantages. All systems function in the same manner, regardless of their physical differences.

QUESTIONS

1. What is meant by the term *amplifier*?

2. Name two characteristics of an electrical wave.

3. What is a block diagram?

4. How is the block diagram used by a technician who must do repair work on an electronic device?

5. What purpose is served by a power source?

6. How does a stereo amplifier-system differ from a monaural amplifier-system?

7. Name three tape recording systems.

8. What advantage does using the cartridge or cassette have over using a reel-to-reel system?

9. What does the term *oscillator* mean?

10. What is meant by the term *bias*?

11. What is the difference between a tape player and a tape recorder?

Block Diagrams: Radio Receivers

4

The radio frequency spectrum includes a number of broadcast services. All broadcast stations in the world are assigned their operating frequencies by their governments. International governmental agreements have designated specific portions of the radio frequency spectrum for specific uses on a worldwide basis. These agreements have given individual governments a great deal of freedom in dividing broadcast frequencies in order to meet the needs of their own nations. In the United States, as in most nations, all broadcast stations are licensed by an agency of the federal government. This agency, the Federal Communications Commission, examines all license applications, sets up frequencies or bands of frequencies for specific use, issues licenses, and polices the actual broadcasts to see that the licensees do not violate established rules and regulations. In the United States, broadcast bands have been established for home-entertainment use. The four basic bands are the AM radio band, the FM radio band, VHF television, which covers channels 2 through 13, and UHF television, which includes channels 14 through 83. The AM radio band has been assigned frequencies from 540 kilohertz to 1.6 megahertz (540,000 hertz to 1,600,000 hertz). A higher set of frequencies, 88 to 108 megahertz, has been established for the FM broadcast band. All radio broadcast stations are assigned their operating frequencies within these bands by the Federal Communications Commission. This permits many stations to broadcast at the same time without interfering with each other.

Radio receivers are designed to meet certain electronic criteria. Among these are three basic considerations: sensitivity, selectivity, and fidelity. *Sensitivity* is the ability of the receiver to *pick up,* or receive, the broadcast signal. Normally, the further away from the broadcast station, the weaker the received signal will be. A receiver with high sensitivity is able to receive very weak broadcast signals. *Selectivity* means that the receiver is able to tune in one station and reject all others. Two stations are often received at once when tuning with low quality receiver, and the listener's efforts to separate the stations often result in his hearing only annoying squeals and howling sounds in the background. Quality receivers are designed to overcome this situation. The third consideration for a receiver is fidelity. *Fidelity* is the ability of a receiver to faithfully reproduce the transmitted signal, with little or no loss of what has been broadcast. It would be fairly easy to place price tags on these three receiver criteria. Generally speaking, the greater the sensitivity, selectivity, and fidelity, the greater the cost of the receiver will be, with very few exceptions. As new electronic devices are invented, this may change.

This chapter covers radio-system block diagrams. The three most common home-entertainment radio systems are discussed: AM, FM, and FM Stereo. Both the similarities and the differences among these systems are presented.

4-1 Transmission of Intelligence

Five things are necessary before transmission of a message from one place to another can occur. This is true regardless of the means used to carry the information. Necessary are: a transmitter, a receiver, a common language, a means of carrying the information, and the specific information itself. These five items are necessary in all cases. Someone or something must be able to initiate the transmission. It could be a person writing a message on a piece of paper, or it could also be a television broadcast station. Another individual has to be able to receive the message from the person sending it. Communication occurs when one person sends another person some message, and the receiver receives and understands the message. In order to be able to understand the message, the receiver must understand the language that the sender uses. The message itself also must be understood. It would not be effective to send a message in code, using words both persons understood, if the person receiving the message did not know what the code was.

The final item required for good communication is something to carry the message. Here again, it may be a piece of paper upon which the message has been written, or it may be an electronic signal from a broadcast transmitter. Typical radio and television entertainment-broadcasting systems use an electronic signal called a *carrier* to carry the intelligence from the broadcast station to the receiver. The intelligence sent out from the station is added to the carrier by a unit called a *modulator.* There are different systems of modulating a carrier that are used by home-entertainment broadcast stations. Two of the most common systems are *amplitude modulation* (AM) and *frequency modulation* (FM). To put it simply, AM broadcasting causes a change in the amplitude of the carrier signal. This amplitude change is directly related to the shape of the electronic signal representing the intelligence. FM systems cause a slight change in the frequency of the broadcast station's carrier in order to transmit the intelligence. Both systems have advantages and both are in use by the broadcast industry. Radio broadcast stations use either AM or FM systems. Television uses both systems, AM for picture information and FM for the sound information. Both of these systems are discussed in detail in later chapters of this book.

4-2
The AM Radio

One of the early kinds of transmission of radio signals used the AM signal. In this kind of broadcasting, the amplitude of the transmitted signal is varied in step with the information being broadcast by the station. Development of this type of radio has evolved over the years into a very sophisticated unit called a superheterodyne radio. The superheterodyne radio is made up of several distinctive functional blocks. These blocks are similar in function regardless of the components used in the radio. Both transistor solid-state units and vacuum-tube units use the same functional block diagram. Electronic signals received from the transmitting station are processed in the same way through the radio, regardless of whether it uses transistors, integrated circuits, or vacuum tubes to do the work of converting the broadcast signal into something that is heard by the human ear.

The AM radio is divided into functional blocks, as shown in Fig. 4-1. The signal from the transmitter is *picked up* by the antenna and then

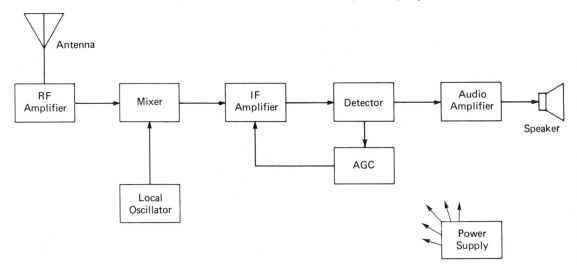

Figure 4-1. The AM-radio block diagram. The blocks found in this radio are found in most radios produced in the United States.

processed through the set to the speaker. Customarily, both block diagrams and electronic-schematic diagrams for home-entertainment devices display the input of the unit on the left side of the drawing and the output on the right side. (This information can be of great assistance to the person learning to work with these diagrams.)

Antenna. Technically speaking, the antenna is considered a part of the radio-frequency amplifier section of a receiver. However, since a radio may have an antenna other than the one built into the set, it has been included here as a separate element. The purpose of the antenna is to pick up a broadcast station's signal and bring this signal into the receiver for processing into sound waves.

RF amplifier. The *radio-frequency* amplifier does exactly as its name implies. It is an amplifier designed to work in the radio frequency range. It provides for tuning or selecting a specific frequency and also amplifies the signal received on this frequency. The output of this block is sent to the mixer block.

Local oscillator. This oscillator produces an electronic signal of its own which is sent to the mixer block. This signal is also tunable. The purpose of this block is to provide an electronic signal, generated in the radio receiver, which will mix or blend with the signal selected by the tuning elements in the radio-frequency amplifier. Most radios produced in recent years are designed to tune both the RF amplifier and the local oscillator by means of turning a single control knob.

Mixer. The mixer also is named well. The signals it receives from the RF amplifier and local oscillator are blended electronically. One of the results of this action is a third signal, called an IF, or intermediate frequency, signal. This signal has a frequency of 455 kilohertz for an AM radio. The signal processed by the IF amplifier block contains the information as received from the broadcast station and another signal, called a carrier, operating at a resultant frequency of 455 kilohertz.

IF amplifier. Here, too, the function of the block is described in the name. The new frequency produced by the mixer action (455 kilohertz) is called an *intermediate frequency*. It is neither the incoming signal from the broadcast station, nor is it the local oscillator frequency. Remember, an amplifier will amplify that which it is designed to amplify, in this case, the IF frequency.

Detector. The signal processed so far in the radio consists of two parts. One part is the music or voice information sent from the transmitter. The other part is called the carrier. The purpose of the carrier is to carry the information from the transmitter to the receiver, at a specific frequency. Each broadcast station has its own carrier frequency.

Once the carrier gets the information to the receiver, it, the carrier, is processed along with the information through the mixer. The mixer operation produces another carrier—at the IF frequency. This new carrier contains the information sent out on the broadcast station carrier. This IF carrier and its intelligence are sent through the IF amplifier to the detector. After this point, the carrier is no longer needed and, for all purposes, it is ignored by the rest of the set. The detector separates the information from the carrier and sends the information to the audio amplifier.

Audio amplifier. The audio amplifier in a radio operates in the same manner as it does in other systems. The input to the audio amplifier is designed to accept a small electrical signal. The electronics contained in this block enlarges the signal and provides sufficient power to operate a loudspeaker. Actually, the audio amplifier block can usually be subdivided into several blocks. These blocks could be identified as a preamplifier, a voltage amplifier and a power amplifier.

Speaker. The speaker, as an output transducer, changes the electrical waves into sound waves. These waves can be heard by the ear. This completes the journey of the information from the broadcast station to those wishing to receive it.

Power supply. All electronic devices have specific operating voltage and current requirements. The power supply is designed to provide these from whatever external source, whether battery or power line, available. The specific amounts of voltage and current are determined by the engineer or designer of the set. They will vary according to exact requirements of each set.

Automatic gain control. The automatic gain control, or *AGC*, section of a receiver is utilized to compensate for different signal strength levels received from the transmitter. This system monitors the amplitude of the signal at the detector. It then returns a part of the detected signal to the IF amplifier and RF amplifier in order to control the amount of amplification of these stages. Usually, if the received signal is strong, the AGC action will reduce the amount of amplification in the RF and IF amplifiers. On the other hand, low signals will allow the AGC action to increase amplification occurring in these stages.

4-3
The FM Radio

One of the reasons for the development of FM broadcast was the desire to eliminate atmospheric noise from being processed by the receiver. This noise sound becomes the static heard on the AM receiver. It is very unpleasant to hear. The quality of reception is better without it, of course.

The AM receiver has two faults which could not be overcome. These faults are poor fidelity reception and an inability to eliminate static-like noises from the reception. These two drawbacks were overcome with the development of the FM transmitter and receiver. Atmospheric noises, such as static, tend to modulate the amplitude of the waveform sent from the transmitter. Using a system that is able to limit the size (height) of the waveform used eliminates these kinds of signal interference in the receiver.

The problem of fidelity is not due to a fault in the AM receiver. It has been caused by an overcrowding condition permitted in the AM broadcast band. The large number of AM broadcast stations licensed to operate in this band has been the problem. Each broadcast station is limited to the amount of space it is able to use in the broadcast band. This limitation does not permit the full range of audio frequencies to be broadcast by the station. FM broadcast stations do not have this kind of limitation. They have the ability to broadcast almost the full range of audio signals. In fact, some FM stations broadcast frequencies much higher than the 20 hertz to 20 kilohertz audio signal. FM stereo broadcasters transmit modulated signals, with information, going as high as 75 kilohertz on the carrier. While enjoying the ability to transmit a full-fidelity and static-free signal, FM does have limitations. One such limitation is that the broadcast station signal is restricted in range to line-of-sight reception.

Now take a look at the FM-receiver block diagram, shown in Fig. 4-2. Keep in mind that we are not looking at the electronics involved in the radio at this time. We are just looking at the block diagram and the basic function of each block. Look for similarities in function between the AM radio and the FM radio while reading through this section. These similarities are discussed at the end of the chapter in a comparison of the two systems.

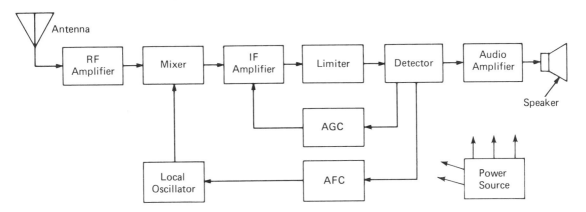

Figure 4-2. The FM-radio block diagram. Two blocks have been added to this radio, in addition to those used in the AM radio.

Antenna. This antenna serves the same purpose as all other antennas. It receives the broadcast signal and sends it to the RF amplifier. The set circuitry will be different in an FM receiver because of the difference in frequencies being processed. A detailed discussion of this is given in a later chapter.

RF amplifier. The RF amplifier performs two operations in an FM radio. These are to select, or tune, the incoming signal and to amplify this signal to a level that permits the rest of the receiver to process it. The RF amplifier acts in the FM receiver in exactly the same manner as an RF amplifier operates in the AM receiver. There are basic circuit differences due to the differences in the frequencies being received. The function of this block is identical in the AM radio and the FM radio; the frequency being processed is different.

Mixer. The mixer takes the received signal containing a carrier and the information from the RF amplifier. It also takes a signal from the local oscillator and mixes or heterodynes these signals. The result of the mixer action is the production of a third signal, called an intermediate frequency signal. This signal is made up of an IF carrier, operating on a frequency of 10.7 megahertz, and the intelligence found on the broadcast station carrier.

Local oscillator. The local oscillator generates an electronic signal. This signal is tunable. The oscillator signal is tuned at the same rate as the incoming signal in the RF amplifier by means of a tuning control that uses a single knob. The purpose of the local oscillator is to generate a carrier signal and to inject this signal into the mixer block in order to have the two signals which are required to produce a third signal, the IF signal.

IF amplifier. An intermediate frequency signal of 10.7 megahertz is produced in the mixer stage. It is made up of the intelligence received from the broadcast station and a carrier. The IF amplifier amplifies this signal from its low level signal at the mixer output. It is amplified through several stages in order to obtain a large signal for the detector stage.

Limiter. A limiter stage is used in the FM receiver between the intermediate-frequency amplifier and the detector. The function of the limiter is to present a signal of constant amplitude to the detector. In this way, any undesired information, such as static, on the signal is eliminated before it reaches the detector. Otherwise, these unwanted signals would be heard as extraneous sounds or as audio distortion if permitted to be detected. Some electronic detector circuits are not bothered by variations in the amplitude of the FM signal. Sets using these detector circuits may not have a true limiter stage. Most sets, however, utilize a limiter to ensure quality reproduction of the broadcast intelligence.

Detector. The purpose of the detector block is to extract the broadcast information from the IF signal's carrier. At this point in the receiver, there is no further need for any carrier. It has accomplished its mission of bringing the broadcast information through the receiver, in an amplified form, and presenting this information to the detector. After the information is removed from the carrier by the action of the detector, the information is sent to three other blocks in the set. Each of these blocks is important to the overall operation of the set. They are the audio amplifier, the automatic frequency control, and the automatic gain control blocks.

Audio amplifier. The audio amplifier receives the information from the detector. The overall purpose of the FM receiver is to reproduce information, in the audio range, as broadcast by the transmitting station. The audio amplifier enlarges the information. It may use several stages of amplification in order to produce a signal that is large enough to be processed by a power amplifier. After the power amplifier stage, the information is sent to a loudspeaker. In the loudspeaker the electrical signal information is converted back into sound waves so that it may be heard.

Automatic gain control. Automatic gain control (*AGC*) systems are used in receivers in order to maintain a strong signal at the input of the detector block. The AGC system takes a sample, or small percentage, of the signal processed in the detector and returns this sample to the IF amplifier. In some sets this AGC signal is returned to the RF amplifier too. The sample signal is used to control the amount of amplification in the IF and RF blocks. It acts just like an automatic volume level control. If the amplitude of the signal at the detector drops, then the AGC action causes the IF and RF blocks to increase the amount of amplification. In this manner the AGC block attempts to maintain a constant amplitude of signal input to the detector block.

Automatic frequency control. The purpose of the *AFC*, or automatic frequency control, block is to keep the radio receiver tuned to a specific station. It is bothersome to have to frequently adjust the tuning control knob on a receiver whenever the station seems to fade out. In many cases this problem is not the fault of the broadcast station. It often originates in the receiver. When the receiver is on, some of its parts change their values due to the heat generated by the operation of the set. This heat may cause de-tuning of the signal being received. The block most affected by changing values caused by heat is the local oscillator. A small signal from the detector is returned to the proper portion of the local oscillator. This signal is used as a correcting signal. It tells the local oscillator that it has changed its frequency a small amount. The signal also attempts to correct this frequency change. It is capable of controlling or holding the local oscillator on frequency if the shift is not too large. When large changes occur, then the listener must provide this correcting action by readjusting the receiver controls.

Power source.   The power source is one of the most important blocks in any receiver. If there is a failure in this block, whole sections of the set may not function. In the case of a major power source failure, the entire set may not operate. The purpose of this block is to provide the proper operating voltages and currents for the operation of the set.

4-4
FM Stereo

The FM stereo signal is processed in the same manner as the FM monaural signal. This was one of the requirements specified by the Federal Communications Commission, when it approved a system for broadcast of FM stereo signals. The system had to be compatible with standard FM broadcasting. Non-stereo receivers had to be able to receive the FM stereo signal and reproduce it as a monaural signal. Stereo receivers had to be capable of receiving a non-stereo signal and reproducing it as a monaural signal. How this is accomplished is covered in detail in Chapter 9, which discusses electronic signals in radio receivers.

Looking at the block diagrams, in Figs. 4-2 and 4-3, for the two FM receiver systems, we see that the blocks are identical from the antenna to the detector. However, in the stereo receiver, once the signals pass through the detector, they go to a new kind of block. This block is called a stereo decoder block.

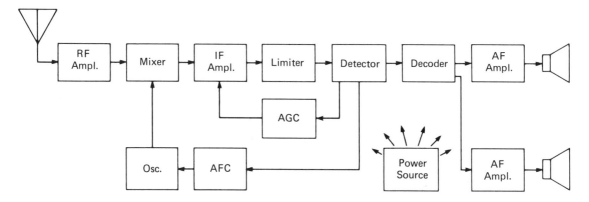

Figure 4-3. The FM-stereo radio block-diagram. Addition of a stereo decorder and another amplifier channel make this unit different from the monaural FM radio.

The stereo decoder block.   The stereo decoder block performs several functions in FM stereo and radio receiver. One of its simplest jobs is to operate a stereo-indicator lamp when a stereo signal is being received. One of its more complex functions is to separate the right channel and left channel audio information and send each channel's information to its corresponding audio amplifier and loudspeaker.

The stereo decoder block is capable of operating in a non-stereo mode. If the incoming signal does not contain stereo information, then the decoder passes the signal on to the output circuit in the decoder. This output circuit then sends the same audio information to both audio amplifiers in the system.

The stereo decoder is actually made up of several sub-blocks. These are shown in Fig. 4-4. In brief, these blocks function in the following manner:

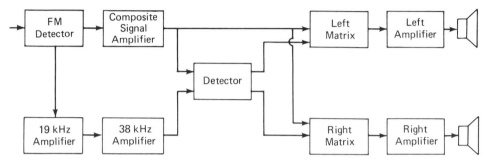

Figure 4-4. The FM-stereo decoder block-diagram. Stereo information is processed by the decoder's blocks. Monaural information bypasses the decoder section. Both signals meet at the matrix, where they combine to produce right- and left-channel information.

19-kilohertz pilot amplifier. The stereo signal broadcast from the transmitter consists of many kinds of information, some of which are not heard by the human ear. One such signal is the 19-kilohertz pilot signal. This signal is used to turn on the FM stereo decoder section of the receiver. The 19-kilohertz signal is then fed into a frequency-multiplying circuit which doubles this frequency to 38 kilohertz. Included in this block is a limiter circuit which limits the amplitude of the incoming signal.

Detector. The FM stereo information is detected from the 38-kilohertz carrier. This is accomplished in the detector block. The details of this process are discussed in Chap. 9. The output of the stereo demodulator (the detector) is sent to the matrix block.

Composite signal amplifier. The composite-signal amplifier amplifies the monaural FM signal. This amplified signal is made up of both right- and left-channel information. The output of the composite-signal amplifier is then fed into the matrix block in order to construct the right- and left-channel information.

Matrix. A matrix is a device that has the capability to mix, or blend, electronic signals. In the FM stereo receiver, the matrix is used to reconstruct right- and left-channel information. It receives signals from the detector and from the signal amplifier and combines them. This results in the right- and left-channel audio signals, which are then sent to their corresponding audio amplifiers.

The audio amplifier. There are two identical audio-amplifier blocks used in a stereo receiver. They are usually identified as the right-channel amplifier and the left-channel amplifier. Each amplifier input is wired to a corresponding output-connection at the stereo matrix block.

Power source. The power source block provides the necessary operating voltages and current to the receiver. There is one common power source for most FM stereo receivers. It provides the correct values of voltage and current for the receiver section as well as for the audio amplifiers found in the radio.

4-5
Similarities
between AM and
FM Receivers

Refer again to Fig. 4-1 and Fig. 4-2. There is little difference in the block diagram of an AM receiver and of a monaural FM receiver *as far as the basic block diagram is concerned.* There are great differences in electronic circuitry between the two radios. These differences are discussed in later chapters. Compare the input to each radio. Both have antennas, RF amplifiers, mixers, and local oscillators. These blocks may be generally classified as the *tuner* section of the radio. The purpose of the tuner is to select one broadcast station composite-signal, consisting of the carrier and the information, amplify it, mix it with the local oscillator signal and, as a final result, present an IF signal to the IF amplifier. Here, too, the composite signal is amplified and sent on to the detector. In both the AM and FM detector, the information is removed from the carrier, and then sent on to be amplified and converted to sounds by the loudspeaker. There are few differences between the block diagrams of the AM and the FM monaural radio. As previously stated, there are major differences in electronic circuitry, but we are not looking at this aspect at this time.

4-6
Differences
between AM and
FM Receivers

The basic differences between these two radio systems, excluding FM stereo, are in the frequencies received and the kinds of detection systems used. One other difference is that the FM radio requires an automatic frequency control (AFC) system to stabilize the local oscillator and thus keep the station tuned in.

FM stereo requires some additional blocks. A stereo amplifier-system is required in order to reproduce a stereo signal of any kind. Also included are two speaker systems, one for each channel. Another item of major importance is a decoder block to separate the stereo information into right and left channels.

As you study the basic block systems of radios and TV sets, look for blocks that are common to all of the systems. Understanding the purpose, or functions, of these blocks will help you when you are servicing these units and will make overall understanding easier. Understanding how the blocks work and, later on, how electronic signals are processed through each of the blocks, as well as what electronic signals are processed, will aid anyone attempting to repair home-entertainment devices. These same principles may be applied to other functional units, including cars and home appliances. Learn the system. As you study this book, you will find that your increased understanding of the block systems will make most repairs easier to diagnose.

SUMMARY   Three types of radio receiver systems are in use in the United States. They are AM radio, FM radio, and FM stereo radio. Each of these systems receives the electrical signals transmitted from the broadcast station and processes the signal in order to reproduce the broadcast information.

Each radio receiver system uses a similar block diagram. There are some additional blocks required for FM and FM stereo radios. Other than the additional blocks, the functional blocks in all three radio systems are very similar, both in arrangement and purpose.

QUESTIONS   1.   What are the four home-entertainment broadcast bands used in the United States?

2.   What frequencies are in use for the AM broadcast band?

3.   What frequencies are in use for the FM broadcast band?

4.   State three criteria which must be met in the design of a radio.

5.   What work is performed by an automatic gain control system?

6.   What advantage does FM reception offer over AM reception?

7.   What work is performed by the automatic frequency control system?

8.   What blocks are used by both the AM and FM radios?

9.   What blocks are added for the FM radio?

10. What additional blocks are required in order to receive a stereo FM signal?

Block Diagrams: Television Receivers

5

Let us consider the nature of the input and the outputs as they relate to a television set. On the input side of the large block we call the TV set, there is an antenna. This is shown in Fig. 5-1. On the other side, there are two outputs, different from the other, but both outputs, and necessary for the operation of the set. One of these is the audio output. A loudspeaker in the set acts as a transducer to convert electrical waves into sound waves. The second output is the picture. A cathode-ray tube is used to convert another group of electrical signals into the visual image displayed on the face of the CRT. Both outputs are required to be in working order for the TV set to function properly.

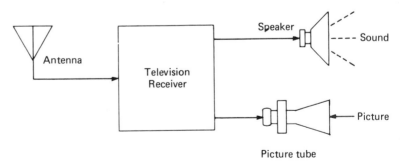

Figure 5-1. A block diagram for a TV set. There is one signal-input at the antenna for this system. The loudspeaker and the cathode-ray tube are the two output devices.

The basic television set must do three things correctly. These are:

1. Reproduce a picture of the proper brightness and color.
2. Reproduce the proper sound to go with the picture.
3. Synchronize both picture and sound with the TV station transmission.

In discussing the cathode-ray tube, in Chap. 2, we said that the electronic beam in the CRT is able to move both across the face of the tube and up and down as well. In order that a good picture be reproduced, the timing of this movement must be synchronized with the transmitted picture signal. This is accomplished by adding synchronized signals to the picture information sent by the broadcast station.

Breakdown of the basic block identified as a TV set will isolate the systems performing the three functions, listed above, into blocks. The system block-diagram shown in Fig. 5-2 illustrates how the signals from the broadcast station are processed in the television set. It shows how the

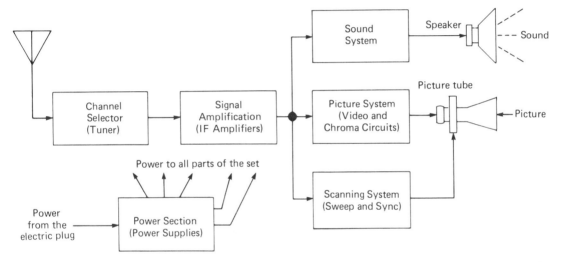

Figure 5-2. A system block-diagram for a TV set. These blocks process all video and sound information, as transmitted. In the center, the signals split into sound, video, and scanning blocks.

three basic functions of sound, picture, and synchronization are related and how they are processed inside the set. Even though they are eventually separated, they must all work at the same time. The signal from the antenna is processed by the tuner. The tuner selects the proper channel from the 12 VHF and the 70 UHF channels available. The received information is then amplified in the IF strip. After this, the three basic signals from the broadcast station are separated and sent to the systems that will further amplify and process them.

These three systems operate almost independently of each other. If one of the systems fails, the other two may still work properly in the set. As the repair technician becomes familiar with the operation of each block in the system he becomes able to relate set troubles to specific blocks. In other words, if one of the blocks were to fail, the technician could use his knowledge of how the set should work and be able to relate the symptoms of failure to that one block or to a group of related blocks in the set. He should be able to spend less time on the actual repair than he would if he did not use this knowledge of the function of each block in the set, and of how each block is dependent upon other blocks, to aid him in his work.

5-1
Monochrome
Receivers

Once the television signal processing-system is understood, we can move on to a study of the building blocks of the television set. Fig. 5-3 shows the block diagram of a typical monochrome-television set. As the function of each block is described, note where the block inputs and outputs are shown on the diagram. This will help you understand how the three basic signal-systems perform their jobs.

Antenna. Here, as in the radio, the antenna is not truly a separate block. Since it is often externally connected to the set, and the set may not work well without the external connection, it should be considered. The purpose of the antenna is to receive the broadcast station signal and send it to the RF amplifier in the tuner.

RF amplifier. The RF amplifier performs two functions in the TV set. One of these is to select the proper broadcast station frequency. Once the proper station is *tuned in,* the received signals are amplified in order to make them large enough to be processed by the mixer. A second input to the RF amplifier is from the AGC block. This will be covered later.

Local oscillator. The local oscillator generates its own electronic signal. The channel selector device usually is designed in such a manner that both the RF amplifier and local oscillator frequencies are selected at the same time. Turning the channel selector knob on the set does this. The output of the local oscillator block is sent to the mixer block.

Mixer. The mixer block takes the signal from the local oscillator and from the RF amplifier and mixes these two signals electronically. The result of this action in the mixer is the production of a third signal. This signal is called the intermediate-frequency signal. It consists of a carrier, on a frequency of 45.75 megahertz, and the information broadcast from the television station. The intermediate-frequency signal is sent to the IF amplifier for further processing.

Tuner. In many TV sets, the three blocks just described—the RF amplifier, the mixer, and the local oscillator—are contained in a separate

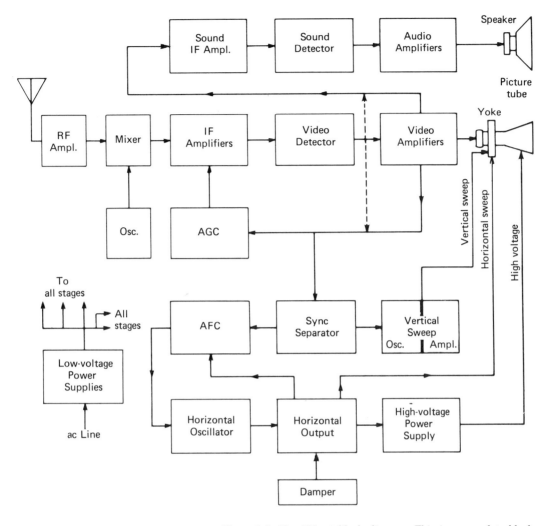

Figure 5-3. The TV-set block diagram. This is a complete block diagram for all functional blocks in the black-and-white TV. Connections are shown by lines, with arrowheads indicating direction of signal processing.

unit in the cabinet, called a tuner. Because of this physical separation from the main chassis of the set, and because all station tuning is done in these blocks, these blocks are often combined into one larger block. This block is referred to as the *tuner* of the set.

IF amplifiers. The signal from the tuner block is not strong enough by itself to operate the sound, picture, and synchronizing systems of the TV set. The IF amplifier enlarges the IF carrier and the transmitted intelligence to whatever predetermined levels are required to operate the balance of the set. This block has to amplify all of those signals necessary to the reproduction of the transmitted signal. This includes the picture information, the sound information, and the synchronizing signals. Most of the signal amplification in the TV set occurs in the IF amplifier block. It is not unusual to have an amplification factor of 8,000 times the size of the signal received from the mixer block. There is more than one IF frequency in use in a TV set. The main IF is 45.75 megahertz. This is the frequency for the picture-information carrier. Another frequency is 41.25 megahertz. This is the sound IF frequency in the main IF block. When color signals are broadcast, a third frequency is used for the color information. This is 42.17 megahertz. All three IF frequencies are used by the color TV receiver. Each is discussed in detail later in the chapter covering signal flow in the TV set.

Video detector. The amplified IF signal is now large enough to be distributed to three main sections of the set. These sections are the sound section, the video section, and the sweep section. The purpose of the video detector is to remove the sound, video, and synchronizing information from the IF carrier. Once the information has been detected, the original 45.75 megahertz IF is no longer required. Part of the detection process produces another IF carrier-signal. This signal is at a frequency of 4.5 megahertz and is the sound IF carrier. The sound information is sent on this new, 4.5 megahertz, IF carrier to the sound section of the TV set.

A second function of the video detector block is to feed back a portion of the detected signal to the AGC block. A third function is to send the detected signal to the video amplifier. The final function is to send the synchronizing signals to the sync-separator block. In some sets, the outputs for sound, AGC, and sync may be connected to the video-amplifier block. The exact connection would be an engineering or design decision. The technician refers to the schematic diagram for the specific set in order to determine where these outputs are located. The function of the output is the same regardless of how it is connected in the set.

Video amplifiers. The video, or picture, information has to be further amplified in the set. This is accomplished in the video-amplifier block. Signal amplification is necessary in order to send sufficient information to the TV picture tube where it is changed from the electronic signal, processed by the TV set, into a visual image we are able to view.

Automatic gain control. So that good picture quality is maintained, the amount of signal amplification in the set will vary according to the strength of the broadcast station signal. If the received signal is weak, the set will have to amplify it more than it would a strong signal. A portion of the video signal is taken from either the detector or the video-amplifier block and fed back to the AGC block. In this block the amplitude of the signal is used to control the amount of amplification in the IF-amplifier and the RF-amplifier blocks. Most TV sets have an AGC adjustment control somewhere on the chassis. This control is considered to be a secondary adjustment, to be made by a qualified technician. When the AGC is properly adjusted, the picture quality is excellent, with good contrast between black and white levels and enough video information so that the picture does not appear faded or *washed-out*. The AGC block adjusts the amount of amplification in the IF amplifier and RF amplifier automatically in order to maintain high standards of picture quality.

Picture tube (CRT). The output of the video amplifier is sent to the picture tube. The electron beam, as it moves across the inside of the face of the tube, translates the video signal into a picture. The beam of electrons moves in two directions; starting at the top of the screen, it moves horizontally across the tube and vertically down, at a fixed rate of speed. The amount of electron-beam energy will vary in proportion to the picture's lightness or darkness, thus, an electronic picture is painted on the face of the CRT. In the United States the television picture is produced by use of 525 individual lines of information. In other words, the electron beam moves across, or scans, the face of the CRT 525 times in order to produce one picture. The beam's movement is designed to sweep all of the odd-numbered horizontal lines on one vertical coverage and then to return to sweep all of the even-numbered lines, thus skipping every other line. There are 30 sweeps a second for each group of lines on the picture. This indicates that the beam is moving at a rate of 15,750 times a second (30 × 525 = 15,750). There are two sets of horizontal lines which make up the complete picture. Each set appears 30 times a second for a total of 60 times (30 × 2 = 60) during the standard time interval of one second. These frequencies of 15,750 hertz and 60 hertz are standard frequencies used in all TV sets produced for sale in the United States.

The visual-output transducer of the TV set is the picture tube. There are four inputs to this tube. In order to have a functioning output, all must be working correctly. These four inputs are the high voltage, the vertical sweep, the horizontal sweep, and the video (picture) information. Two of the inputs are connected electronically to the tube. These are the video and the high voltage. The magnetic fields developed in the deflection yoke are the other two inputs to the CRT. Even though these two inputs are not wired directly to the tube, they have a direct influence on the movement of the electron beam and must be considered a part of the CRT input.

Sound IF amplifier. The sound information is sent out by the transmitter along with the picture and synchronizing information. This signal has its own IF frequency. When the sound signal is processed in the TV set, both the sound IF carrier and the information on this carrier are amplified.

The frequency of the sound IF carrier is 4.5 megahertz. The sound is broadcast on an FM carrier and must be amplified to a high enough level in order for detector action to occur. This amplification of the sound and its FM carrier is the function of the sound IF block.

Sound detector. The sound detector separates the sound information from the sound carrier. It then sends this information to the audio amplifiers. This block functions in a manner similar to other detectors.

Audio amplifiers. The sound information is amplified in this block. The information from the detector is not large enough to operate the loudspeaker. The purpose of the audio amplifier block is to amplify the sound signal so that it is powerful enough to make the loudspeaker work.

Loudspeaker. The loudspeaker, or speaker, translates the sound information into audible sound. While this unit is often considered to be a part of the audio amplifier system, it should not be overlooked when troubleshooting.

Sync separator. One of the many outputs of the video-detector or video-amplifier blocks is a group of electronic signals called synchronizing pulses. There are two sets of these pulses sent by the transmitter with each set of picture information. These pulses are designed to keep the receiver's picture in step with the transmitter's picture. One set of pulses contains synchronizing information for the vertical sweep system. In the set, the vertical sync pulses tell the vertical sweep system when to start its scanning pattern. The other set of sync pulses transmitted in a non-color broadcast are the horizontal sync pulses. They are designed to tell the horizontal sweep system when to start its scanning. Both sets of sync pulses are sent to a sync-separator block. Electronic circuits in this block separate the pulses into those that are processed by the vertical system and those that are processed by the horizontal system. Without these sync signals the picture will either roll vertically or pull horizontally. In some cases both will occur.

Vertical sweep system. The vertical sweep system is made up of at least two blocks. These blocks are the vertical oscillator and the vertical-output amplifier. The purpose of the vertical sweep system is to force the electron beam in the CRT to move up and down the face of the CRT at a controlled rate. This controlled movement means that the beam has to move vertically 525 times in order to complete one picture. Some block diagrams show the vertical section as one block, identified as the vertical-sweep block. However, what occurs in this one block can be better understood if the block is thought of as two major sub-blocks.

Vertical oscillator. The input block in this system is the vertical oscillator. This electronic-signal generator produces a sawtooth-shaped waveform. A synchronizing pulse from the sync separator provides a signal telling the oscillator when to start. The purpose of this oscillator is to generate a kind of signal that is able to change at a steady rate. This sawtooth signal is then sent to the vertical output block.

Vertical output. The vertical output block amplifies the signal received from the vertical oscillator. This action is necessary in order to produce a signal that is powerful enough to develop changing magnetic fields in an output device called a *deflection yoke*. This yoke is not considered to be a separate block in the set. It acts as an output device for both the vertical-sweep and the horizontal-sweep sections of a TV set.

Horizontal sweep system. The purpose of this system is to provide sufficient signal to the horizontal windings of the deflection yoke. The blocks for this system are very similar to those of the vertical sweep system. These blocks are identified as the horizontal oscillator, the horizontal output, the damper, and the AFC block. Note that this section requires two more blocks than does the vertical sweep section of the TV set. Let's look at all of these blocks in order to see how they function in the set.

Horizontal oscillator. The purpose of the horizontal oscillator block is to produce an electronic signal that, when further amplified, will cause the electron beam in the CRT to scan the tube horizontally. The shape of the electrical signal produced in this block is a sawtooth waveform. A signal from the AFC block is provided as a synchronizing signal, telling this oscillator when to start.

AFC. The AFC block is a frequency-controlling block. The initials stand for *automatic frequency control*. Its function is to see that the horizontal-oscillator frequency remains constant at the proper rate of 15,750 hertz. This block has two inputs. One consists of the sync pulses from the sync separator. The other input is a pulse fed back from the horizontal output stage. These two inputs are compared in an electronic circuit and frequency changes are then corrected in the same circuit. The output of this block is a signal, at the proper operating frequency, that tells the horizontal oscillator when to start. This signal keeps the oscillator frequency correct.

Horizontal output. The signal from the horizontal oscillator is sent to the horizontal output block. Amplification occurs in this block in order to make the signal powerful enough to control the magnetic fields of the horizontal windings on the deflection yoke. This signal causes the magnetic fields of the horizontal windings of the yoke to change intensity at a controlled rate. The electron beam sweeps across the screen as the fields change.

Damper. The function of the damper is to control any undesired electronic-signal generations which occur in the horizontal-output circuit due to the higher frequencies used in this block. Damping means to *reduce vibrations*. In the case of the TV set, these vibrations are unwanted electronic signals produced in the horizontal circuits.

High-voltage power supply. One output of the horizontal circuit sends a signal to the deflection yoke. Another signal is fed back to the AFC block. A third signal goes to the high-voltage power supply block. High voltage is developed in this block in order to provide a charge on the CRT for picture brightness. This high voltage is developed during the *retrace* period, after the electron beam has completed its movement across the face of the CRT and is returning to the left side in order to sweep across the tube again. The high voltage is produced when the beam is quickly returning. Because of this rapid voltage development during the retrace time, the high-voltage power supply is called a *flyback* supply.

Low-voltage power supply. Any electronic device used for home-entertainment purposes requires a source of power to operate it. The power may be obtained from batteries or it may come from a local power-generating company, when the set is plugged into a wall-outlet. The low-voltage power supply block in the TV set is designed to convert the incoming electrical power into the operating voltages and currents required for set operation. *Low voltages* in a TV set are defined as any of the operating voltages other than the high voltage needed for picture tube *raster* operation. (Raster, as used in a television set, means the formation of an illuminated screen due to the scanning process. Picture information does not have to be present in the raster.) These operating voltages range from low values of six volts, used to heat the filaments in vacuum tubes, to a high value of 400 or 500 volts, required for electron flow in vacuum tubes. Solid-state devices, such as transistors or integrated circuits, require voltage values between five and 160 volts. Engineering design determines the exact voltage and current requirements for each TV set.

Almost every electronic device produced has an area on its chassis that contains the components used for power supply. Newer techniques in set design have modified this location. More recently designed sets use a traditional power supply in order to obtain directly only some of the required voltages. Other voltages are obtained from blocks which have not been identified, or thought of, as power sources. For instance, some TV sets develop low voltages from the sweep, or scanning, blocks. These scan-derived power sources serve as fail-safe devices. (Without them, failure of the sweep circuit could cause overheating of the scanning output components and possibly a fire in the set.) If the scanning circuit is not functioning properly, then the power source in it will not develop the proper operating voltages for the balance of the set. Most schematic diagrams used by technicians identify all of the power sources in the set. Use of this information when servicing the set reduces the location of all power sources to a relatively simple operation.

**5-2
Color
Receivers**

A key word in TV broadcasting is *compatibility*. This is the ability of a set to receive both black-and-white and color broadcasts from the TV station. Compatibility was one of the major concerns of the Federal Communications Commission and the electronics industry when a usable color-system was being developed. The TV broadcast station adds color information to the station carrier when it broadcasts a color picture. This color information is added in such a manner that it does not interfere with non-color reception. The transmitted color signal may be divided into two major segments. One of these is the color-synchronizing signal and the other is the color, or *chroma,* information. Color-processing blocks are added to the basic black-and-white TV set when a color set is produced. Figure 5-4 shows how the color information is taken from the video circuits in order to be processed by the set. Two signal paths are used—one for the video information, and one for the color information. The color information is processed in the set and then sent to the picture tube, where it is mixed with the picture information. If color information is not being broadcast, the color section of the set does not turn on, and the black-and-white picture information shows on the CRT. When the color information is processed, three color-signals are produced. They are red, blue, and green. These are the three basic colors used in a color-TV set. Other colors are reproduced by blending some of the basic colors together.

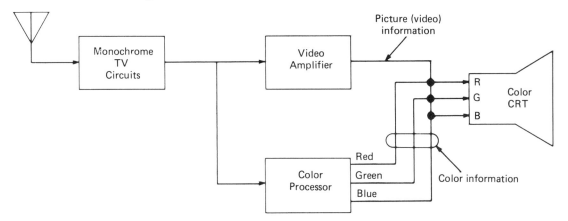

Figure 5-4. A color-system block diagram. This expands the color processing system to show that the video information is split away from the color information. The video and color signals are recombined at the CRT.

The block diagram for a color TV shown in Fig. 5-5 displays the same blocks found in a black-and-white TV set *and* the additional color-processing blocks. There is no need to describe the non-color blocks again, since they have just been explained. Instead, let us see what occurs in the color-processing blocks. These blocks are displayed in the lower

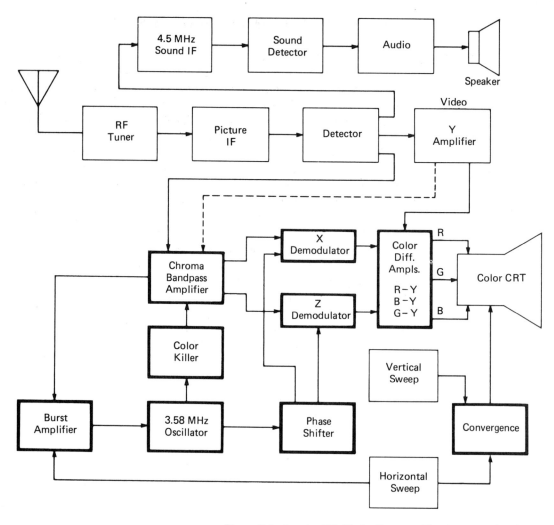

Figure 5-5. A color-TV block diagram. The boxes outlined by heavier lines contain all the functional blocks directly related to color processing. All other blocks are the same as used in a black-and-white TV set.

portion of Fig. 5-5. Circuitry will vary among makes, but basic functions remain the same.

Color sync. Synchronizing signals are transmitted with black-and-white picture information in order to stabilize, or *lock in,* the vertical and horizontal signals in the receiver. These same signals are transmitted with a color broadcast in order to do the same job. Color information also must be synchronized if the receiver is going to do its job of faithfully reproducing the color picture. The color-sync signal rides along with the other sync pulses until it reaches the color-sync block in the set. In this block, it is separated from the other information coming from the video circuits and sent on to other color blocks.

Color oscillator. The color-information signal sent from the transmitter is a special kind of signal. It is unusual in that the carrier used has been reduced to a very low amplitude. In order to detect or demodulate the color information, this color carrier must be re-created in the receiver. The purpose of the color oscillator is to generate a color-signal carrier of 3.58 megahertz in the receiver.

Color killer. The color-killer block is used as a circuit-switch in the TV set. If there is a color signal and if the color oscillator is operating, then the color-killer block is turned off by the color signal. However, if there is no color information being received, the color-killer block remains on. If this killer circuit is on, then the balance of the color circuits are inoperative. Non-color signals are then processed as normal black-and-white picture information. The basic function of the color-killer block is to turn off the color blocks in the set when the signal being received is not a color signal.

Color IF. The color IF-amplifier block amplifies the color-signal information sent from the video section of the TV set. This information is often called chroma information. Also, the block itself is often called the *chroma bandpass amplifier*. Acting in a manner similar to the video IF-amplifier, this amplifier enlarges the chroma information to a relatively high level so that detection action may occur in the next block.

Demodulators. The demodulator block removes the color information from the color carrier. In a system called suppressed-carrier modulation, the color information is added to the main transmitter-carrier. This system, which is explained in detail in Chapter 10, requires that a carrier be added to the signal in the receiver before detector action occurs. Figure 5-5 shows the 3.58-megahertz oscillator-output signal entering two demodulator blocks. It also shows the signal from the color IF amplifier entering the same two demodulator blocks. These two signals entering each demodulator block reconstruct a modulated carrier. Detection, or demodulation, occurs in these blocks. The outputs of these two demodulator blocks are used to reconstruct the red, green, and blue colors used in the receiver.

Color amplifiers. The three basic color-signals require further amplification before they are strong enough to be used by the color picture-tube. In addition to this amplification, the picture information from the set's video amplifier must be combined with the color information in order to faithfully reproduce the transmitted picture-information signal. This is accomplished in the color video-amplifiers and the information is then sent to the picture tube for display. When all of these signals arrive at the CRT at the proper time, a quality picture is presented.

**5-3
Comparison
of Systems**

In discussing the audio amplifier, AM and FM radio, and monochrome TV, so far, the similarities between the types of sets have been brought to your attention. Also, detailed block diagrams of each of these types of units have been presented. We have seen that functional blocks with the same operational title perform the same kind of work in all kinds of home-entertainment electronic equipment. For instance, an audio amplifier amplifies audio signals, regardless of the kind of set using it. If you keep this idea in mind as you become familiar with these various blocks, learning how each type of set works will be made easier. A comparison of color TV and black-and-white TV shows that the only basic difference is in those blocks which process the color information. The blocks which process black-and-white information perform the same function in both types of sets.

There are basic similarities in *function* among all kinds of receivers. The frequencies of the signals received from the broadcast stations will be different, and the complexity of the signals will vary, but each type of set has basic block similarities which, when understood, will help you understand how any set functions.

Antenna. In order to receive the broadcast signals, the set requires some kind of antenna system. This may be a built-in device, or it may be a very sophisticated outdoor unit.

Tuner. A tuner is made up of three basic blocks. These are the RF amplifier, the mixer, and the local oscillator. One purpose of the tuner section is to select the proper station and amplify it. Another purpose is to mix the RF signal and a signal from the local oscillator in order to create a modulated IF frequency for use in the IF amplifier.

IF amplifiers. Each type of receiver has its own IF frequency. These frequencies have been standardized and are the same regardless of the make of the set. The IF amplifiers further amplify the signal so that it may be further processed in the set.

Demodulators. Each type of set separates the signal information from the IF carrier in order to further process the signal. There are different kinds of demodulators, but their purpose is always the same.

Power supply. All sets require some kind of power source. It may be a battery, or the set may be connected to a wall-outlet. The power supply section changes the power from whatever source into the specific operating voltages required for each set.

Audio amplifiers. When sound is a part of the input to the set, this sound is amplified and reproduced by the loudspeaker. This is true of all systems reviewed so far including the audio amplifier.

Picture systems. Both black-and-white and color TV sets have video systems. These systems amplify the picture information. Once this is done, the video information is sent to the CRT for display.

Scanning systems. The requirements for the movement of the electron beam in the CRT are similar in color and in black-and-white TV. The beam is moved (swept) across the face of the CRT as well as from top to bottom of the screen. These movements are synchronized by the sync section.

Color section. This section is used only in a color TV. It is turned off by the color TV set when a black-and-white picture is received from the broadcast station; it is turned on when a color signal is received. It is not included in a non-color set.

SUMMARY   There are only a few basic building blocks used by home-entertainment electronic devices. Figure 5-6 shows the major blocks in a

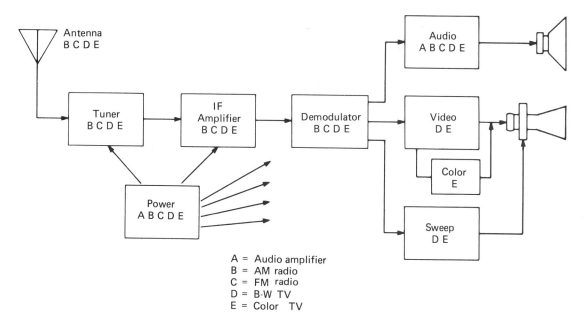

Figure 5-6. This is a major-block diagram for a color TV set. Coding in the diagram indicates similarities among blocks used for other systems discussed.

color TV set. Letters in each block indicate in which other systems the same type of block is to be found. Notice that many of these blocks are used in the majority of the types of units. Only the video, color, and sweep systems are not used by almost all of the systems reviewed.

Comparing the systems clearly shows how each system uses similar blocks in order to process the signals as received at the input of the set. Analyzing the types of sets into their systems shows that there really are just a few kinds of systems and that the *function* of each similarly named block in the compared systems is very similar. This knowledge may be applied to any kind of system block diagram in order to help understand how the signals are processed through the system. Chapters 7 through 10 cover in detail how these signals are processed in each of the sets we have reviewed.

QUESTIONS

1. What input transducer is used for a TV set?

2. What output transducers are used for a TV set?

3. Name the three requirements for reproduction of a television signal.

4. What three functions are processed by a TV set?

5. What is meant by the term *sync*?

6. What purpose is served by the vertical sweep system?

7. What purpose is served by the horizontal sweep system?

8. How does a deflection yoke work?

9. What blocks are used in a color TV that are not used in a black-and-white TV?

10. List all of the blocks that are common to radio and TV receivers.

Block Diagrams: Basic Measuring Equipment

6

Diagnostic tools are as important to the electronic technician as they are to the technician in any field. The technician uses basic measuring devices such as multimeters or oscilloscopes in order to measure actual working values or signals in a set. These pieces of testing equipment help the technician in his analysis of problems. He probably will start the repair process by turning on the set and watching, listening, or smelling for trouble. After an initial diagnosis as to which blocks of the set are malfunctioning, he will then take out the manufacturer's service literature to look at a schematic of the set. Good schematic diagrams have both operating-voltage values and typical waveforms shown on them. Good schematic diagrams also show exactly how the set is wired and locate test points at which signals are to be measured.

Using test equipment, the technician makes the measurements of voltage, current, resistance, or signal waveforms and compares these values to those indicated in the service literature. When he discovers some values that do not agree with the service information, he has probably located the trouble area in the set. Then the technician uses his knowledge of how the circuit is supposed to work and the information obtained by use of test equipment in order to locate a specific part that has malfunctioned. Once the bad part has been replaced, the technician checks circuit values again in order to confirm that the repaired set works according to the manufacturer's specifications.

This chapter is devoted to block diagrams of measuring equipment.

There are many varieties of basic test and measuring instruments on the market today. A good technician knows his tools and how to get the most out of them. Test equipment is certainly considered to be in the tool category. Occasionally, a technician may be able to diagnose a fault and locate a defective part in the set simply because he has done the same kind of analysis on similar sets many other times. Analysis by experience is good in some cases, but it is possible to be in error and to get started on the wrong path because of this kind of repair technique. A much better approach is to use diagnostic tools in order to locate a fault. The block diagram approach shows you how each type of test unit is put together. Each piece of equipment has limitations. Knowing how the test equipment works, when to use a specific unit, and any effect the equipment has on the set when its used to make a measurement helps the repair technician make his analysis work much easier.

6-1
The Basic Tester

Circuit values that are commonly measured in electronic work are the voltage, the current, and the resistance. In some instances these are also related to time. In other cases only an instantaneous value is important. Here is one place where selection of the proper instrument for the test is critical. Before we look at specific kinds of testers, let's look at the basic test system used. Figure 6-1 shows a typical system, with both an input and an output. Most measuring devices use wires for connecting the set's

Figure 6-1. A basic tester is made up of input connectors, called test leads, the tester and its circuitry, and some kind of output-indicator device.

test points to the tester. These may be considered as the input to the tester. In some cases these wires have handles, or holders, attached to them for the technician's use. Test wires such as these are called test *leads*. Note that the tester has two leads. This is a requirement because all testers measure some quantity *with respect to a specific point in the set*. When a voltage is measured, it is accepted practice to state, for instance, that "there are 140 volts at test-point A." What is really being said is that an electric potential of 140 volts exists between point A and a common reference point in the set. This common reference point is often called the *ground* or *common*. In actual practice, this is the electrical path or point which is used as a reference for the set. Many sets use the metal chassis itself for this point.

6-2
The Basic Meter

In this chapter, we are discussing how basic test and measuring equipment functions. Each test unit takes a circuit value and processes it in the tester. There are several types of output transducers in common use with measuring devices today. The most common ones until recent years have been the *analog meter* and the *cathode-ray tube.* A fairly new addition to the readout family is the *digital readout.* Photographs of an analog meter and a digital readout are shown in Fig. 6-2.

Figure 6-2. Two types of meters. Shown above is an analog readout device. On the left is a digital readout, as used with test equipment. (Courtesy Simpson Electric Co.)

A meter may be used for a specific application, such as monitoring a power line voltage. In such a case, the technician uses a voltmeter that is capable of reading voltages in the area of 120 volts. The meter has to be able to measure values both above the 120 volt mark and below this value in order to give a true reading of the measured voltage. Most meters available today measure from zero to some specific value, such as 150 volts. Here, the technician refers to the meter as having a *range* of 150 volts. He assumes that everyone knows that the meter starts at zero. The block diagram for the basic meter is very simple. It consists of the input leads and the meter movement (the output unit), as shown in Fig. 6-3.

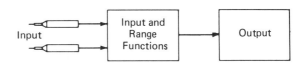

Figure 6-3. This basic tester is similar to that of Fig. 6-1 except that the tester has been identified as a meter.

The home-entertainment repair technician has to measure many quantities. It is much more convenient to use a multipurpose test meter than it is to have to connect several separate test meters to the circuit. Testers that are capable of measuring voltage, current, and resistance are in general use today. These multimeters, as they are commonly called, are designed to select one of these three basic circuit values by simply moving a switch in the input circuit from one position to another on the meter case. In addition, there is another switch which is capable of extending the measurable value range of the meter. This provides additional adaptability for the user. These meters measure voltage, resistance, and current. They are commonly referred to as *VOM's* or volt-ohm-milliammeters since these are the quantities that they are designed to measure. The output of the multimeter consists of one meter face with the meter face showing several scales. There is at least one scale for each function of the meter. Often there is more than one scale for the same function, depending on how the meter is designed. Figure 6-4 shows pictures of multimeters in current use.

Figure 6-4. Two popular types of multimeters used by technicians. (Courtesy B&K Precision Dynascan Corp.)

6-3
Electronic
Multimeters

The electronic multimeter was developed in order to overcome some limitations in the standard multimeter. One such limitation is a condition called *loading.* This refers to the effect the meter has on a circuit when the meter is used to measure a circuit value. Under the proper set of circumstances the addition of a meter to a circuit can provide a reading error of over 50% of the actual values in the circuit. This is because the test meter introduces a certain amount of additional resistance into the circuit. Even if the test meter's resistance is fairly low, this will affect circuit operation. Better test meters have a very high internal resistance in order to minimize this effect.

Looking at a block diagram of the electronic multimeter, Fig. 6-5(a),

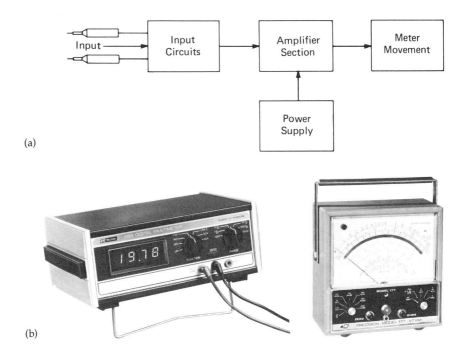

Figure 6-5. (a) The block diagram for an electronic multimeter. Note the addition of an amplifier and a power supply. (b) Two popular electronic multimeters—one analog and one digital. (Courtesy B&K Precision Dynascan Corp.)

we see that two new blocks are shown. These blocks consist of an amplifier and a power source. The amplifier works in a similar manner to those used in audio circuits. It amplifies the electric signals as received from the input selector. The power source is necessary for operation of the amplifier circuits. It may be required for the output circuits as well.

There are two kinds of output meters generally used with electronic multimeters. One is the analog-meter movement, as used with nonelectronic multimeters. The other readout device is a digital readout —additional circuitry in the meter converts the electrical signals into a digital form which is then displayed by the readout. Typical electronic multimeters used by electronic technicians are shown in Fig. 6-5(b).

The meter, by its basic design, is only able to measure values as they occur in a set. The technician compares these measured values with the values given in the service literature for the set in order to determine if the set has the proper operating voltages, currents, and resistances. There comes a time when the repair technician needs more information than is obtained by looking at a meter scale and reading the indicated value. This occurs when the technician needs to know the form of the electrical signals, as well as their values. He then uses another kind of tester called the oscilloscope.

The oscilloscope, shown in Fig. 6-6, adds one more dimension to a measurement. It is capable both of displaying amplitude, or strength, of the signal and of relating this to a time reference. With the dimensions of strength and time being displayed, the oscilloscope is actually drawing a graph for the observer.

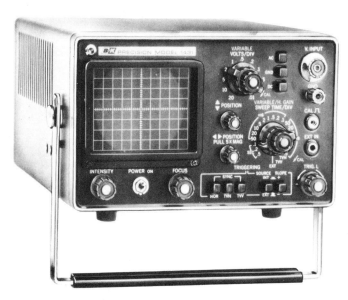

Figure 6-6. Oscilloscope used by electronic service technicians. (Courtesy B&K Precision Dynascan Corp.)

Electrical signals used in home-entertainment devices have certain forms, or signatures. The manufacturer's service literature shows both the shape of the signal and its amplitude. The oscilloscope is used to display the actual signal present in the set. This is then compared to the service information. If the two do not agree, then a trouble area in the set has been located. The technician then moves to adjacent areas in order to find exactly where the trouble originates. This approach is called *signal tracing*. It is a very effective method of troubleshooting a set. Look at the block diagram of an oscilloscope in Fig. 6-7. Several of these blocks are similar to others investigated so far.

The vertical system. The *input selector* processes the input signal and sends this signal to the vertical amplifier. The *vertical amplifier* enlarges the signal so that it can be displayed on the face of the CRT. As the name suggests, the vertical system controls only that portion of the signal which is in a vertical position—in other words, the height, or amplitude, of the display.

Inside the CRT are two pairs of metal beam-deflecting plates. The output of the vertical amplifier is a voltage which is connected to these

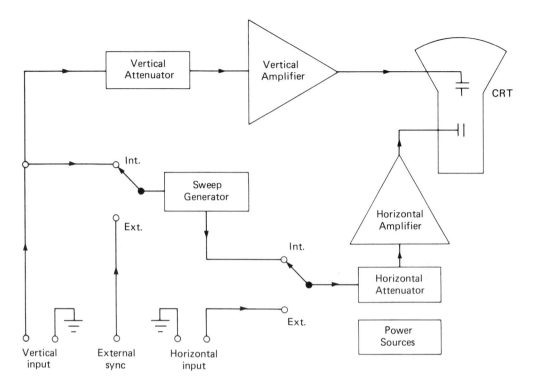

Figure 6-7. The block diagram for an oscilloscope. Through use of switches, various horizontal inputs are selected. Sync signals may also be selected by a switch.

vertical deflector plates. When the voltage on these plates is equal, or balanced, the electron beam is centered on the screen of the CRT. As the signals processed by the oscilloscope increase in amplitude, one of the deflection plates gets a higher voltage on it. The electron beam, being negatively charged, is drawn toward the plate with the higher voltage. The result of this action causes the displayed beam to move up on the face of the CRT. If the signal voltage is of a lower value than the *no-signal input* condition where the beam is centered on the screen, then the output of the vertical amplifier will cause the beam to deflect downwards in the CRT. Translating all of this into other words, the electron beam in the CRT is normally centered when displayed on the screen of the CRT. A voltage, or signal, applied to the vertical input of the oscilloscope will cause the beam to move, or deflect. When a positive voltage is applied, the display moves up. When a negative voltage is applied, the display moves down. A constantly changing input voltage or signal will cause the display to be constantly moving as it shows the values of the input on the face of the CRT.

Another part of the vertical sweep system is the *vertical attenuator.* Attenuate means to reduce. A vertical attenuator reduces the amount of the voltage applied to the vertical amplifiers of the oscilloscope. The amplifiers of the oscilloscope are designed to provide a specific amount of deflection with a predetermined input voltage to the amplifier. As an example, let's assume that an input voltage of 0.1 volt causes the beam to deflect one centimeter on the face of the tube. If the CRT face is ten centimeters high, then an input voltage of 0.5 volts will make the beam move from center screen to the top or bottom of the face of the CRT. This amount of deflection sensitivity limits the range of the oscilloscope. In order to extend the range of voltages that the instrument is able to measure, an attenuator is included between the input connection of the oscilloscope and the vertical amplifier. This input attenuator consists of a network of resistances that are designed to reduce the input voltage to an amount which provides full deflection of the electron beam. Most oscilloscope attenuators are calibrated so that the operator is able to determine the input voltage by looking at the calibration markings on the front panel of the oscilloscope. A front panel range-marking of five volts per centimeter and a display with an amplitude of three centimeters indicates a measured value of 15 volts.

The horizontal system. There is one other pair of deflecting plates inside the CRT. These are placed in such a manner that voltage variations on them will cause the beam to move horizontally. You may recall that the oscilloscope is designed to display waveforms with reference to time. The *horizontal deflection plates* are usually connected to a time-base generator. The purpose of this generator is to develop a varying voltage which, when applied to the horizontal plates, will cause the beam to move across the face of the CRT in a linear manner (that is, at a constant speed). The speed of horizontal movement is called the time-base of this generator. Combining vertical deflection and horizontal deflection voltages results in a display that faithfully reproduces the input signal waveform, as shown in Fig. 6-8. The amplitude of the time-base voltage is controlled so

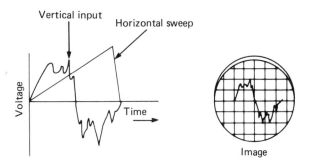

Figure 6-8. The oscilloscope displays amplitude variations on the vertical scale and relates these variations to a time reference. The resulting image is the display seen by the observer.

that the deflection is across the face of the CRT. There are times when the technician wishes to compare a vertical signal with a horizontal signal. To do this, the time-base generator is turned off and a system similar to the vertical input is employed. This is illustrated in Fig. 6-7, also. A switch connects a horizontal amplifier and its input system to the horizontal deflection plates instead of the time-base generator.

Figure 6-7 also shows an attenuator for the horizontal amplifier system. This *horizontal attenuator* operates in the same manner as the vertical attenuator. It is possible to measure two voltages with this oscilloscope. One voltage is introduced into the vertical system and another voltage is introduced into the horizontal-sweep system. Each amplifier causes the beam to deflect from the center of the screen of the CRT. By use of two deflecting systems, the beam can be placed anywhere on the CRT screen. The amount of beam deflection is directly related to the result of the two deflecting forces produced by the vertical and horizontal amplifiers.

Additional systems. Additional circuitry in the oscilloscope includes a synchronizing section and a power source. The sync section *locks* the display in order to stabilize it for easier viewing. The signals used to synchronize the waveforms may be taken from the vertical input as shown in Fig. 6-7. It is possible to use another synchronizing signal from an outside source as well. Provision is made for this on most oscilloscopes by means of a selector switch on the front panel. Most oscilloscopes have two power-source blocks. One is for the high voltage required to get the electron beam to the front of the CRT. The other is for the low voltage supply, and its function is to provide the necessary operating voltages and currents for the proper operation of the amplifiers and time-base generator of the oscilloscope.

Dual-trace oscilloscopes. Synchronization of signals in a television set becomes more important as the complexity of the set increases. It is often necessary to observe two signals at the same time in order to determine if they are synchronized. Figure 6-9 shows the block diagram of a dual-trace oscilloscope. Another vertical sweep system has been added. The CRT has two vertical amplifier inputs. Circuitry in the scope sets up two displays, one for each input system. There is only one set of horizontal plates, so the horizontal-deflection rate is the same for both vertical signals. Using two vertical inputs, two different signals may be displayed at the same time. Each vertical amplifier has its own positioning control so that either the two displays may be separated from each other or one display may be positioned on the other one for a close comparison of timing and shape. The blocks in a dual-trace oscilloscope are identical to those of a single trace oscilloscope except for the addition of a second vertical-sweep system.

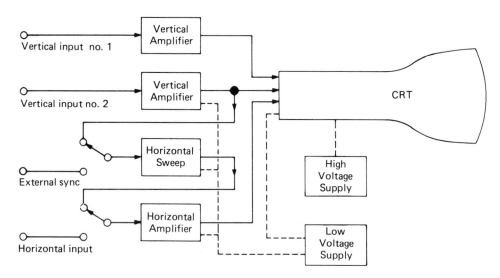

Figure 6-9. A dual-trace oscilloscope uses two vertical input systems in order to display two waveforms on the CRT at the same time.

6-5
What Is Measured

The study of electricity or electronics is complete only when all the various terms used are completely understood. Certain of these terms have been used by this author recurrently so far in this book. Some have been defined. Others have been given a hint of definition. Basic terms such as *resistance, current, voltage,* and *power* are in daily use by technicians. Definitions for these terms are sometimes hard to express. Confusion over terminology can be a problem among newcomers to the field of electronics. Sorting out the differences between these terms and their possible interrelationships at this point will help the reader understand the balance of the book.

The basic electric circuit. The terms about to be defined can have little meaning until the framework for using them can be related to some known object. The simplest electrical object in which work is done is the basic electric circuit. The circuit consists of three major items. Each item has its own identity. Each is a part of the whole. Without all of them the circuit will not operate. They are independent things, but they work only as an interdependent group.

One of these three items is the source of electrical energy. This *power source* develops the energy to meet the specific requirements of the rest of the circuit. The exact makeup of the source is not important at this time.

A second item in our electrical circuit is the load. The term *load* represents the work being done. One of the simplest loads is an electric light bulb. The load may be as complex as a color television set. Here, as with the source, the internal workings are not important except as a means of knowing what specific power values are needed in order to operate the load correctly.

The third important item in an electrical circuit consists of the lines or wires which connect the source to the load. Without these lines, there would be no work done. Both the source and the load would be ready to do some work, but until the lines are connected nothing can be accomplished. Keeping the basic requirements for an electric circuit in mind, let's now explore the specific qualities involved in the circuit.

Current. This book is concerned with the study of the electron at work. Electrons move through conductors, such as a copper wire, at a constant speed of 186,000 miles per second. Since the speed is constant, we must consider the quantity of electrons that are moving. Electric *current* is defined as the number of electrons that move past a given point in an electric circuit in a time period of one second. This current is measured in *amperes*, or amps. When an expression such as "there are 15 amps in this circuit" is used, it really means that a quantity of 15 amperes of electric current are moving past one point in that circuit in one second. In electronic work, terms for smaller quantities of current, such as *milliamp* (1/1,000 of an ampere) and *microamp* (1/1,000,000 of an ampere) are in common use.

Voltage. Water does not normally flow uphill. A pump or pressure system is required before any uphill flow will occur. The same is true in electronics. Some materials, such as copper or silver, have electrons moving about freely in them. An outside force is required to make these electrons move in some specific direction. The force used in electronics is called *electromotive force*, or *EMF*. EMF is the electrical energy required in a circuit to move any quantity of electrons in order to do some work.

Potential energy is energy available to perform some work. Electrical potential is the electrical energy available to move electrons in the circuit from the source to the load and back to the source. In order to move electrons in a circuit, there must be a difference in electrical potential between two points in the circuit. Electrons, being negatively charged particles, move from a field that has a high concentration of electrons to a field that has less of a concentration. A field that has a great many electrons is said to be negatively charged. A field with fewer electrons, than the one used as a reference, is said to be positively charged. In other words, electron movement is from negative to less negative, or positive, in a circuit.

The difference in the electrical charge at two points in a circuit is called the *potential difference* of the circuit. It is measured in volts. When such a statement as "there are 24 volts at that point" is used, the speaker is really saying that there is a difference of 24 volts in potential energy from the point being discussed to a reference point in the circuit. The reference point is often called a *common*, or sometimes a *ground*. Keep in mind that voltage *exists* between two points in any circuit. It is the quality of a circuit that causes electron movement. Another circuit factor which directly affects electron flow is the opposition to movement, or resistance, in the circuit.

Resistance. All materials have an atomic makeup. The electron is part of the atom. Some materials have electrons which are locked into their atoms and are not free to move away. These materials are called insulators. Other materials give up electrons easily. These are called conductors. Between the two extremes of conductors and nonconductors are a large number of materials which permit some electron movement. These materials are classified by their amount of opposition to electron flow. This opposition to electron flow is called *resistance.* The *ohm* is the unit of resistance. The circuit's three qualities of voltage, current, and resistance are interrelated. This relationship was discovered by a German scientist, George S. Ohm. In honor of his research, this relationship is called Ohm's law. Simply stated, Ohm found that one volt is required to cause one ampere of current to flow through a resistance of one ohm. A detailed explanation as well as some applications of this law are discussed in a later chapter.

SUMMARY   Electric meters are used to measure circuit qualities of voltage, current, and resistance. Combination meters called multimeters are in general use by electronic technicians. These multimeters can measure any of these three qualities by the movement of a selector switch. Newer versions of the multimeter use an amplifier circuit inside the meter in order to have less effect on the circuit being measured. Output transducers for multimeters are meter movements. Electronic multimeters may use either a meter or a digital readout.

The oscilloscope is another common test instrument. Its output transducer is a cathode-ray tube. The advantage of the oscilloscope over the meter is that the oscilloscope displays the amplitude of an electrical waveform with relation to a standard time reference. The oscilloscope displays voltage waveforms and does not generally display current or measure resistance.

The three circuit-qualities of resistance, voltage, and current have an interrelationship. Ohm's law shows this relationship by stating that one volt will cause one ampere of current to flow through a resistance of one ohm. Knowing this relationship can be useful in determining circuit values. It is one of the most useful tools available when there is a need to calculate amounts of voltage, current, or resistance.

QUESTIONS   **1.** What work is performed by a meter?

2. How does a multimeter differ from other meters?

3. What two forms of output transducers are used by meters?

4. What does the oscilloscope display that cannot be observed with a meter?

5. What similarity is there between the TV's and the oscilloscope's vertical sweep systems?

6. What parts are in a basic electric circuit?

7. What is the unit of electrical pressure?

8. What is the unit of electrical resistance?

9. What is the unit of electrical current?

10. What relationships are shown in Ohm's law?

Signal
Paths

Signal Paths:

Signals

and

Paths

Much has been written so far about signals. It is time that we stop to analyze just what this signal is. Signalling is a means of communication. Certain American Indian tribes used smoke signals in order to communicate over long distances. They were limited to using these signals during daylight hours, and the receiver had to be within sight of the sender's signals. As the United States grew, telegraph wires were strung across the continent. The previously stated limitations were no longer problems. The telegraph operator sent the message by pressing a key (switch). Varying the time the key was held down varied the time the electric circuit was complete. This was translated into the familiar dots and dashes of the Morse code. Thus, the key operator sent a series of electric pulses, which represented a message, through a wire, to the receiver. This is shown in

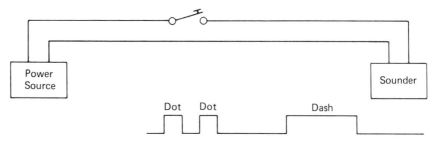

Figure 7-1. A simple circuit, for sending telegraph signals. The lower drawing shows the variation in voltage levels and how they are used to represent dots and dashes for code transmission.

Fig. 7-1. As the telegraph key is depressed, the circuit is completed. A rise in operating voltage causes electric current to move in the wires. This makes the sounder work. Changing the duration of the time the key is held down produces a series of pulses representing dots and dashes.

These electric pulses represent a very simple form of electric signal. The pulse has three important factors which need to be considered when referring to an electric signal. One factor is *amplitude,* the second is *time,* and the third is *shape.* Amplitude refers to the height, or magnitude, of the signal. Time in this case means the number of repetitions of the pulse in the standard time-reference of one second.

What is meant by the term *signal*? In electronics, the signal is the carrier of a message, or intelligence, in electronic form. These signals can have various forms. A person studying electronics has to distinguish between electronic signals and operating voltages when working on sets. Both are present in order to have the set work properly. Both operating voltages and signal voltages are discussed in this book. In traditional books written about electronics, signals are discussed only in conjunction with theory on how circuits function. The author believes that an understanding of what the information-carrying signals are and what they look like is so important that this entire chapter is devoted to this information. The three chapters following discuss how audio, radio, and television sets process these signals, in order to reproduce the information sent out by the broadcast stations.

7-1
Types of Signals

Keep in mind that the simple telegraph circuit described at the beginning of this chapter carries intelligence. The intelligence in this case is conveyed by changes in voltage which represent the dots and dashes associated with the Morse code. This is really an *on-off* kind of circuit. Operating voltages from the power source cause the voltage in the circuit to change from zero, when the key is up, to a constant level, when the key is depressed and the circuit is turned on. If both the sender and the receiver know the same code, then a message is successfully delivered.

Most of the signals used in electronics are more complex than that just discussed. Electronic signals have several characteristics. These characteristics are illustrated in Fig. 7-2. Take a moment to look at this figure. Each of five qualities depicted is characteristic of electrical signals. They are all variations of the three basic waveforms—sine, square, and sawtooth—and are used to develop the shape of the electronic signal used to carry information.

Shape. Each signal has its own *signature*. The shape, or form, is displayed on an oscilloscope in the course of checking out a circuit. Many electronic devices receive a basic signal at their input terminals. These signals are then processed in the set. They may be modified as the processing occurs. The output signal may not look like the input signal. This is not important. The important thing is that the signals located in the

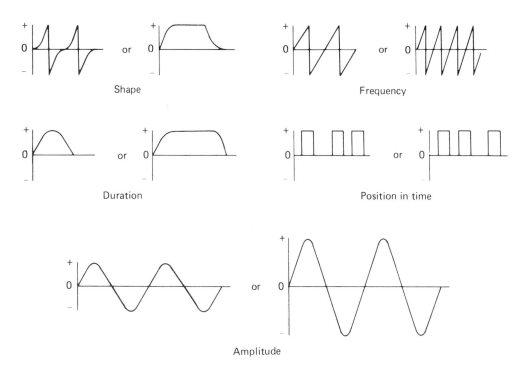

Figure 7-2. Characteristics of electrical signals. Each signal has all of the qualities illustrated.

set by the technician, and displayed on the oscilloscope, have characteristics the same as those called for in the manufacturer's service literature.

Frequency. Each signal has a frequency characteristic. What this means is that the signal repeats a given number of times during the standard time-reference of one second. The shape of the wave does not have anything to do with frequency of the wave. The technician looks for the repetition rate. Earlier in this book, the term for frequency rate was identified as *hertz*. This means cycles per second.

Duration. Another characteristic of the signal is its duration. How long is it on? This is important in circuits that have a controlled time to their operational characteristic. Scanning systems in television sets are on for a specific time period during each cycle. When the circuit is off, the beam returns to its starting point, in order to be ready to repeat the scan.

Position in time. The placement of the signal may be critical during a cycle. Color information in a TV set uses time relationships in order to develop the rainbow of colors displayed on the screen. Position in time is often referred to as the *phase* aspect of the signal. A sine wave reference-signal requires 360 degrees of generator rotation before it repeats. This rotational reference is used to identify timing positions. If two signals start at the same moment in time, they are *in phase* with each other. Should one signal start after the first one, then the second signal is *out of phase* by a specific number of degrees. The number of degrees is determined by how much later the second signal starts. For example, if it started after the first signal had completed 25% of its cycle, then the second signal would be 25% of 360 degrees, or 90 degrees, out of phase with the first signal.

Amplitude. The amplitude of the signal represents how large it is. The amplitude is measured in volts. Usually these measurements are taken from the maximum height to the maximum depth of the signal, that is, from the high peak to the low peak. Since the reference is from peak to peak, then the signal amplitude is measured in peak-to-peak volts. This value may be measured from a zero reference point, as illustrated in Fig. 7-2. It may also be completely unrelated to the zero reference and appear as a changing positive or negative value in the set. There will be more on this later on in the book.

Audio waves. Electronic signals are roughly classified into two groups. One group consists of those signals operating at frequencies in the audio range of 10 hertz to 20 kilohertz. Other signals operating at frequencies higher than those in the audio range are generally classified as radio-frequency signals. This is because signals in this frequency range are transmitted from broadcast stations. All broadcast signals are considered to be in this broad category of radio-frequency signals.

The group of signals classified as audio waves look like any other signals. A pure tone generated in the audio range has a *sine wave* shape. Raising the frequency (pitch) of the tone increases the number of cycles of waves that occur in one second. Reducing the frequency of the tone reduces the number of waves produced in one second. This is shown on the left in Fig. 7-3. The shape of the wave remains the same, but the frequency of the wave increases as the frequency of the tone increases. Waves are often combined into a very complex wave shape as multiple tones, or frequencies, are used. The human voice is a good example of this. The voice consists of several frequencies. A waveform representing this complex signal is shown on the right in Fig. 7-3.

The continuous wave. Once the frequency of a signal becomes higher than those in the group of audio frequencies, this signal may be broadcast from a radio transmitting station. The broadcast station is licensed to use a specific operating frequency. The transmitted signal is made up of component parts, in many cases. What these parts are depends upon the kinds of information being broadcast.

The broadcast signal has to contain one basic unit called the *carrier*. Intelligence may be added to this carrier by a system called modulation. First, let's discuss the carrier. The carrier is a continuous electrical signal

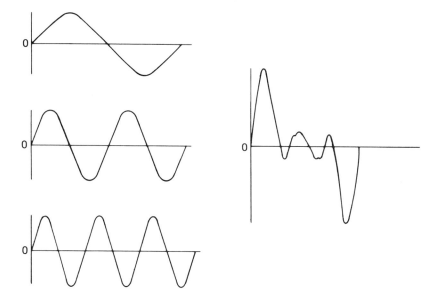

Figure 7-3. A sound wave is composed of several individual waves. Each of the three waves on the left is found in the composite wave on the right.

that uses a specific frequency. Each broadcast station has its own carrier frequency. The waveform of the carrier is a sine waveform. The frequency of the carrier wave depends upon the assigned operating frequency of the broadcast station. This wave is often referred to as a *CW* signal, or *continuous wave* signal. In a simple transmitter system, the carrier is turned on and off in a manner similar to the system used by a telegraph operator. The result of this kind of action is short pulses of carrier signal coming from the transmitting station. The pulses can be used to represent the dots and dashes associated with Morse code.

The modulated carrier. A continuous-wave carrier operating in the radio frequency-range may have lower frequency signals added to it. This process of adding some form of intelligence to a carrier is called *modulation*. The CW signal is modulated by the intelligence. Modulation produces a slight change in the shape of the carrier. Two of the most common modulating systems are amplitude modulation and frequency modulation. There are other modulating systems in current use, but these two are those most frequently encountered by technicians.

The electronic signals used for home-entertainment broadcast purposes are made up of a carrier and modulating information. In some cases, the modulating information contains several different modulating signals. This is true when FM stereo or color television signals are broadcast. Ways have had to be found to add all of the required information to the carrier without distorting the information. Also, ways have had to be developed to remove the information from the carrier successfully. The various methods of putting together a complex modulated signal are discussed in this section. First, we will discuss the basics of amplitude modulation.

7-2
Amplitude
Modulation
(AM)

Amplitude modulation is the process of adding a signal in the audio frequency range to a signal in the radio frequency range. In this process, the audio information is identified as the *modulating signal* and the radio frequency signal is called the carrier. The term *carrier* is easily understood. The carrier is used to carry the intelligence from the transmitter to the receiver. In the amplitude modulation system, the height of the carrier is varied as the intelligence varies. Figure 7-4 shows the two signals and the result of the modulating action on the carrier. The amplitude of the modulated wave retains the form of the modulating signal and varies in step with the shape of the modulation.

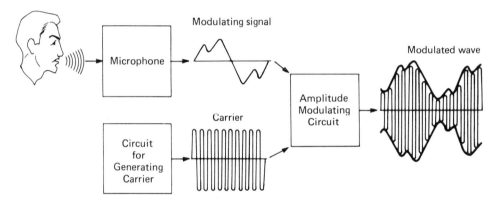

Figure 7-4. The amplitude modulating process. A modulating signal in the audio frequency range is imposed upon an RF carrier signal. The result is the modulated carrier signal.

7-3
Sidebands

Our investigation of the modulation process has identified a signal called a carrier and another signal called the modulation. These combined signals operate at the carrier frequency. However, there are other signals present during the transmission of a modulated carrier. The process of combining two electrical signals produces other signals. Through a process called *heterodyning,* two additional signals are developed. One of these additional signals is the result of adding the carrier and the intelligence frequencies. It is identified as the *upper sideband* frequency. The other signal is the result of subtracting one of the two main signal frequencies from the other one. It is called the *lower sideband* frequency.

These three frequencies—the upper sideband, the original carrier, and the lower sideband—operate together to form the modulated carrier signal. If the carrier is modulated by a single frequency tone, there will be one pair of sidebands created. As the frequency range of the modulating signal varies, there will be one pair of sidebands created for each new frequency.

Bandwidth. As more sidebands are added to the carrier, the modulated signal requires more operating space in the broadcast band. A carrier without intelligence requires a fairly narrow portion of the broadcast band. Assume that we have a carrier operating on a frequency of 800 kilohertz. When this carrier is modulated by a 1,000 hertz tone, the sidebands increase the carrier width to range from 799 kilohertz to 801 kilohertz. Increasing the tone frequency to 10 kilohertz produces sidebands of 790 and 810 kilohertz. The width of the broadcast signal has increased from one frequency to cover a band that is 20 kilohertz wide. The space requirement for the modulated signal is called the *bandwidth* of the signal. It is often shown on a chart called a *spectrum* diagram. One such chart is shown in Fig. 7-5(a). The center carrier is identified as well as both sets of sidebands. Drawing a line around the peaks of these signals, as shown in Fig. 7-5(b), produces a spectrum curve showing the shape of the modulated carrier as it relates to the broadcast band of frequencies. The form of this curve becomes important to the technician when he is working on set alignment and frequency response information. (*Frequency response,* as it is used in electronics, refers to the ability of a device to process a group of frequencies. For example, if an amplifier is able to process those frequencies from 20 hertz to 15,000 hertz without any significant losses, then its frequency response is from 20 hertz to 15,000 hertz. Inability to process specific frequencies would result in a major drop in the shape of the response curve at the specific frequency.) Drawings of this kind are referred to as *bandpass curves,* or *bandwidth curves,* by technicians. They represent the amplitude of the signal at all frequencies used by the signal.

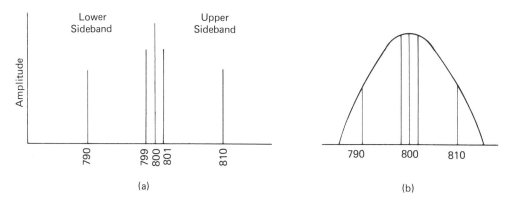

Figure 7-5. Bandwidth of a signal and its sidebands. The sidebands containing information extend above and below the carrier frequency. A spectrum curve is displayed on the right.

Sidebands and sideband modulation. The discussion so far has identified a carrier as made up of a continuous wave. The amplitude of the carrier is determined by the modulation on it. Each tone, or note, of modulating information produces a pair of sidebands identified as the upper sideband and the lower sideband. The combined width of the carrier and its sidebands depends upon the frequencies of the modulation. If the modulating frequencies placed on a carrier were made up of 500 hertz and a 2 kilohertz note then the bandwidth of the station's broadcast signal would be four kilohertz—two kilohertz above the carrier frequency and two kilohertz below the carrier frequency. The frequencies of the modulating information determine the total bandwidth of the signal.

The intelligence contained in an amplitude modulated signal is contained in the sidebands of the signal. The carrier's purpose is to get the message delivered. As a result of the modulation process, there are three signals. These signals are identified as a lower sideband signal, the carrier signal, and an upper sideband signal. Each of the two sideband signals contain the shape of the modulating signal. Both the upper and lower sidebands contain the same intelligence. It would seem that one sideband could be eliminated, and the remaining sideband and the carrier could be used to transmit a message. If this were accomplished, then each broadcast station would occupy less space in the radio frequency spectrum and more stations could be licensed to operate on the airwaves. As things stand now, the Federal Communications Commission has limited the bandwidth of AM broadcast stations to ten kilohertz. This limits the frequency response of the information being broadcast to five kilohertz. By eliminating one of the two sidebands being broadcast, either more stations could operate in the AM broadcast band or the frequency response of each station could be doubled to ten kilohertz.

Sideband modulation variations. Persons working in developmental areas of electronics were able to produce a transmitted signal made up of a carrier and one sideband of intelligence. Experiments with single sideband transmissions were carried out in the early 1930's in the United States. Commercial interests began using this system for overseas broadcasting purposes. It has not been adapted for use in the standard AM broadcast band probably because it is not compatible with standard double-sideband systems.

Using an electronic filtering system, one of the two sidebands is eliminated at the transmitting station. It is almost impossible to eliminate a complete sideband without losing some of the information being broadcast. This is due to limitations in tolerances of the electronic components used in the system. Therefore, there is some overlap in the practical system. This does not harm the quality of transmission or reception in any way.

Several different single-sideband systems have been developed. All

92

use similar methods of transmissions. Each has its own characteristics and should be mentioned. Through the development process, various modulating systems have been used successfully. One system is identified as vestigial-sideband modulation. In this system, the lower sideband is filtered out in the development of the broadcast signal. This leaves the upper carrier and one sideband of information. There is some overlap of signal into the lower sideband due to the inability to eliminate the entire signal without affecting the desired signals. A bandwidth curve for this signal is shown in Fig. 7-6(a).

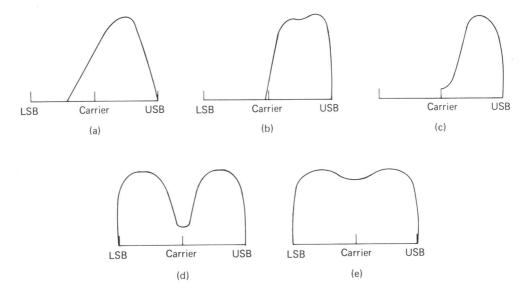

Figure 7-6. Spectrum curves for various AM systems. In order, from left to right, these are: vestigial sideband, single sideband, suppressed-carrier single-sideband, suppressed-carrier double-sideband, and a normal carrier double-sideband signal.

Further research into methods of AM broadcasting found that it is possible to transmit only one sideband of information and to suppress the carrier signal to a very low level. When this process is used, the carrier must be regenerated in the receiver in order to remove the intelligence from it. The advantage of this system is that it takes less space in the spectrum of broadcast frequencies. It also uses less broadcast power since almost all of the transmitter power is now used to broadcast the sideband information and almost none is used to transmit the carrier. The curve for this signal is shown in Fig. 7-6(b).

Two other types of AM signals are shown in Fig. 7-6. One of these is a single sideband signal using a reduced carrier. This reduced carrier in this case, is identified as a *pilot* carrier because it is used as a synchronizing signal in the receiver in order to help in the detection process. In a single-sideband transmitting system the carrier must be re-created in the receiver in order to demodulate the information from the broadcast signal. The phase relation between the broadcast carrier and the regenerated carrier in the receiver is critical. If these two signals are out of phase, then distortion occurs and the quality of reproduction of the signal is poor. In order to keep the relationship of the two signals in phase with each other, a pilot signal is sent from the transmitter along with the sideband information. The curve for this signal is shown in Fig. 7-6(c).

The next curve shown is one for a double-sideband suppressed-carrier signal. There are times when both sidebands are used to transmit all of the required information, but the carrier is not required. A system of this type is used as a part of the FM stereo broadcast signal. Both sidebands are broadcast along with a suppressed-carrier pilot signal. Here, as with the single-sideband suppressed-carrier signal, the pilot carrier is used to synchronize the transmitted carrier with the carrier that is regenerated in the receiver. This curve is shown in Fig. 7-6(d). The remaining curve displayed in Fig. 7-6 is that of a double-sideband full-carrier signal similar to those used in normal amplitude-modulated broadcasting.

7-4 Frequency Modulation (FM)

One of the major disadvantages of amplitude modulation is that electrical discharges in the atmosphere are able to amplitude modulate a carrier. The resulting static and interference in the receiver are annoying to most listeners. The noises often override the modulation, interfering with the broadcast signal. A method in use to overcome this limitation is to modulate the frequency of the broadcast signal instead of the amplitude. When receiving an FM signal, the limiter block in the receiver cuts off any amplitude modulated information. Static and atmospheric noise are not reproduced when this system is used. Figure 7-7 shows how a modulating signal causes a change in the carrier frequency in the FM system. The two signal frequencies are combined in order to produce the modulating effect. In other words, the modulating information causes a slight shift in the frequency of the carrier. As an example, a 10 kilohertz tone modulating the frequency of a 100 megahertz carrier causes a shift in frequency between 100,010 kilohertz and 99,990 kilohertz. Two things occur in a frequency modulation system. One is the amount of change, or deviation, in the carrier frequency. This amount is determined by the amplitude of the modulating signal. The greater the amplitude of the modulation, the greater the deviation from the carrier frequency. The second thing occurring is the rate of deviation. The frequency of the modulation determines the number of times in one second that the frequency changes on the signal. These two quantities—the amount of deviation and the rate of deviation—work together to produce the variations in the carrier that are found on a frequency modulated carrier. These two quantities are used in the FM detector to reproduce the audio information used to modulate the carrier.

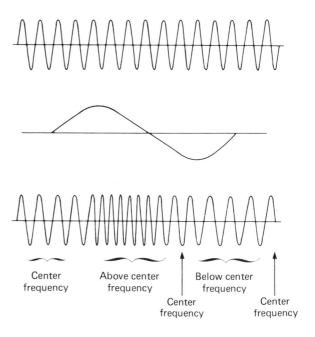

Figure 7-7. Frequency modulation. The audio frequency note is added to the carrier frequency. This produces a deviation in the carrier.

Complex modulation. This chapter deals with the adding of two electronic signals together through a process called modulation. In this process, one signal is the carrier and the second signal is the intelligence. The intelligence modulates the carrier in order to produce what is called a modulated carrier wave. So far, reference has been made to one signal used to modulate one carrier. It is possible to add more than one signal to a single carrier. It is also possible to add more than one carrier onto the main broadcast transmitting-station carrier. It is also possible to add more than one modulated carrier signal onto the main carrier. The process of adding additional channels of information onto one main carrier is called *multiplexing.* Two or more carriers are modulated with their specific information and then these carriers, called, in this case, subcarriers, are used to modulate one common main carrier. The result is a very complex signal containing each modulated subcarrier and the modulated main carrier. In the receiver, this complex signal is passed through electronic filtering devices which separate each subcarrier from the main carrier. After the subcarriers are separated, each is detected in order to remove the specific intelligence contained on the subcarrier. It is possible to transmit and receive several kinds of intelligence at the same time through this system of multiplexing. The next phase of our discussion is to relate this process to practical uses in the home-entertainment field.

FM stereo multiplex. Stereo signals are made up of two or four channels of information. Earlier in this book, we looked at tape recording and
showed how a system is used to reproduce the stereo signal. In radio
broadcasting, similar systems would require the use of two radios in
order to reproduce the stereo sounds. This approach would not be too
practical. It is better to use one radio in order to reproduce the stereo
information. A system was developed, and is now in use, which permits
an FM transmitter to broadcast a monaural signal and a stereo signal at the
same time. This system uses the multiplex process. Let's look at it in order
to see how the signals are put together.

Look at Fig. 7-8. Each channel of the stereo signal is identified. One is
called the L, or left, channel and the other is called the R, or right,
channel. This figure shows the spectrum of an FM stereo broadcast signal
before it is added to the main carrier. The frequencies from 0 to 15
kilohertz contain the non-stereo information. This is identified as the
L+R signal in our discussion. A monaural radio will receive and process
this group of frequencies, ignoring all others that may be on the main
carrier. Another group of frequencies, ranging from 23 to 53 kilohertz,
contains the FM stereo information. This information is contained in a
double-sideband suppressed-subcarrier signal. The suppressed subcar-

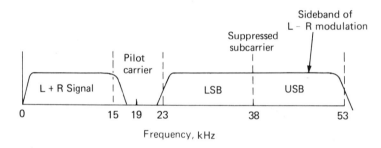

Figure 7-8. The spectrum of the FM stereo signal before it is added to
the carrier. The lower 15-kilohertz signal is used for FM mono and
stereo broadcasting. The upper portion of the signal is used only for
stereo.

rier is centered at 38 kilohertz. Each sideband contains a L−R signal.
The L−R information is amplitude modulated on the 38-kilohertz subcarrier. One other signal appears on the main FM carrier. It is a
19-kilohertz pilot signal. This signal is used in the receiver to synchronize
the regenerated 38-kilohertz signal required in order to demodulate the
suppressed-carrier signal. The reasons for using 19 kilohertz are that no
other information uses this frequency on the broadcast spectrum, and
that the signal can easily be frequency doubled in the receiver to produce
the required 38-kilohertz carrier. The methods of recovering and recon-
structing the FM stereo signal are covered in Chap. 10. The purpose of this
chapter is to show the reader what the electronic signals look like and
what they contain.

The black-and-white TV signal. A black-and-white TV signal contains several kinds of information. Two main items found in the signal are the picture, or video, information and the sound, or audio, information. Video information is transmitted as an AM signal. Sound information is transmitted as a FM signal. Both signals are received at the same time by the TV set. They are amplified in the set in the RF-amplifier and IF-amplifier blocks, and then separated and sent to their specific detector systems in order to extract the information from the carriers.

Synchronizing pulses and color information (when applicable) are transmitted on the carrier in addition to the video and sound information. All of these signals require a large bandwidth in order to successfully transmit the intelligence. In fact, the signals of a TV broadcast station occupy a bandwidth of 6 megahertz (6,000,000 hertz). The spectrum curve for a standard TV channel is shown in Fig. 7-9. The AM portion of

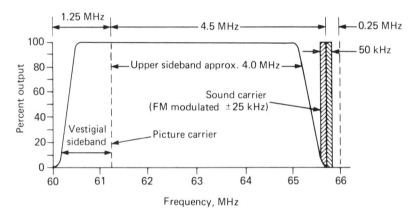

Figure 7-9. The spectrum of a TV signal. This complex signal requires 6 megahertz of bandwidth in order to transmit all of the video, sync, and sound information required to reproduce the picture.

the signal occupies 5 megahertz of the space. The balance is used by the FM audio signal and the AM sidebands. There are standards for broadcasting a TV signal. These are necessary in order to insure that any set produced today will be able to receive a black-and-white TV signal and to reproduce the signal on the CRT and in the speaker. Each TV channel uses 6 megahertz of the spectrum for its signals. The carrier frequency is 1.25 megahertz above the lower band edge, as shown in Fig. 7-9. The sound carrier is 4.5 megahertz above the picture carrier. If a color signal is broadcast, its subcarrier is about 3.58 megahertz above the picture carrier. These separations are always constant regardless of the frequency of the TV station's main carrier. Appendix 1 of this book lists the carrier frequencies in current use for all TV channels. The AM portion of the signal is transmitted in a manner similar to a single-sideband signal. The lower sideband is suppressed, leaving the carrier and the upper sideband. As mentioned earlier, it is impossible to completely remove the lower sideband with this type of transmission. Suppressing one sideband in this manner leaves a portion of the sideband, the vestigial sideband. It occupies some space in the spectrum and is considered a part of the broadcast signal.

The composite video signal.   The FM sound portion of the transmitted TV signal is a monaural signal. It is the same as other FM signals we have covered. There is little need to review it again at this time. Let's look instead at the video portion of the transmitted signal. This signal contains picture information as well as synchronizing signals. The total video signal is called the *composite video signal.*

The video signal is composed of many parts. These parts include the picture information, and synchronizing signals to lock together the transmitter and receiver scanning and blanking pulses that turn off the electron beam as it returns to the left side of the CRT before scanning another line. A composite video signal for three lines of a TV picture is shown in Fig. 7-10. Certain criteria have been established by the broadcast industry relating to the TV signal. Let's look at these before we discuss the composite video signal. There are 525 lines in any TV picture transmitted on the North or South American continents. Each picture is displayed 30 times a second. This develops a scanning frequency rate of 15,750 hertz (30 × 525 = 15,750). The amplitude of the signal is used to control the amount of white or black in the picture. A level of 12.5% of the total amplitude is referred to as the white level of the picture. Black level is at 75% of the maximum amplitude of the signal. Additional information is transmitted above the black level. It cannot be seen by the viewer because that part of the picture is dark. These criteria are also indicated in Fig. 7-10.

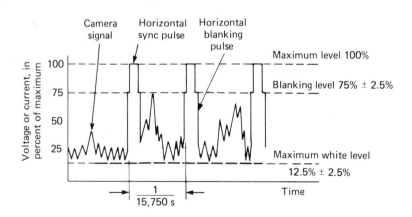

Figure 7-10. Three lines of a video signal. Note the levels of picture information (camera signal) and the relationship to sync and blanking pulse amplitude.

Each line of the TV signal sent from the transmitter contains picture information for that specific line. It also is made up of a horizontal sync pulse and a horizontal blanking pulse. These are repeated at a rate of 15,750 times a second. The blanking pulse turns off the electron beam as it is repositioned to scan another line. The sync pulse synchronizes the beam in the TV set with the beam in the TV camera in order to faithfully reproduce the original picture information. Notice that the sync and blanking pulses operate when the picture is at maximum black level.

The TV picture is produced by a system which interlaces the horizontal lines. This is done in order to produce a clearer picture on the CRT. During one *field* of the picture, all of the even-numbered lines are scanned. During the other *field* all of the odd-numbered lines are scanned. These two scans are repeated 30 times a second. The term used to describe this action is *interlacing*. It refers to the process of interweaving the odd- and even-numbered lines in order to develop a complete picture. One complete picture is called a *frame*. At the end of each field, the vertical blanking pulses are transmitted. These pulses turn off the picture while the beam is retraced to the starting point for the next field. While the beam is retracing and the vertical blanking pulses are being transmitted, vertical sync signals are also transmitted. These pulses ride on top of the vertical blanking signal in a manner similar to the horizontal sync pulse shown in Fig. 7-10. The frequency of the vertical sync and blanking pulses is 60 hertz. The first 21 lines of each field of the picture are used for transmissions of vertical sync and blanking information. This information is shown in Fig. 7-11. If the set's viewer adjusts the vertical control so that

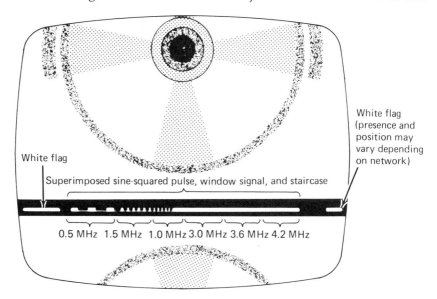

White flag
(presence and
position may
vary depending
on network)

White flag

Superimposed sine-squared pulse, window signal, and staircase

0.5 MHz 1.5 MHz 1.0 MHz 3.0 MHz 3.6 MHz 4.2 MHz

Figure 7-11. The vertical blanking pulse. It uses 21 lines of the picture. Also shown are vertical-interval test signals, at the bottom of the blanking bar.

the picture starts to roll, then these non-picture lines will appear. In addition to the vertical sync and blanking signals, lines 17 and 18 of each field contain information developed at the transmitter and used to provide a standard reference for color and luminance. Other adjacent lines may be used to send out the station call letters or any other pertinent information, such as time or channel number. One reason for development and transmission of a color-standard reference signal is to provide signals at the receiver which will eliminate the need for viewer-adjusted tint or color-intensity controls. Automatic operating circuits in the receiver will adjust for any color or intensity differences, using these signals as a reference. The names for the color test signals are the *vertical-interval test signal (VITS)* and the *vertical-interval reference signal (VIRS)*. Their waveforms, as observed on an oscilloscope, are shown in Fig. 7-12.

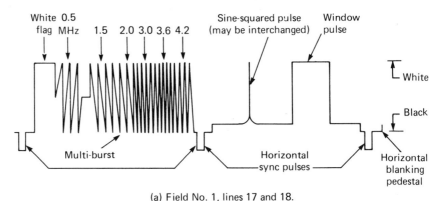

(a) Field No. 1, lines 17 and 18.

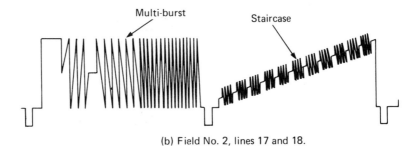

(b) Field No. 2, lines 17 and 18.

Figure 7-12. The composite vertical-interval signals. These are used to check quality of picture reproduction.

The color TV signal. Color TV signals are very complex signals. The color signal contains all information for color reception as well as all of the information necessary to receive a black-and-white picture. All of this information is contained in a signal whose bandwidth remains at 6 megahertz. Much effort has been spent in the development of a color signal for use with home-entertainment receivers. As a result of these efforts, a color TV system has been developed which is compatible with

100

both color and black-and-white reception. Three basic colors are used in order to produce a range of colors in the receiver. These basic colors are red, blue, and green. By use of a system called *color mixing,* these primary colors produce the rainbow of color observed in the TV set. Mixing the colors in the proper quantities will even produce a white image. Let's take a look at all of the components that are required to make up a useful color signal.

The color picture is split into two major sections in order to be compatible for use with a black-and-white TV set. The video information is separated from the color information. The video information is identified as the *Y signal.* Its function is to provide all of the luminance information needed in order to reproduce the black-and-white TV picture. The color information is identified as the *C signal.* It contains all of the chrominance information required to add color to the picture. In addition, there is a color sync signal transmitted. This signal is called the *color burst* signal. It synchronizes the color information in the set with that being transmitted by the broadcast station.

The chrominance signal is made up of two signals identified as the *I* and *Q signals.* These two signals are developed in a way that makes them out of phase with each other. The I and Q signals are used to modulate a subcarrier whose frequency is about 3.58 megahertz. Shifting the phase of one of the modulating signals on this subcarrier permits both of the signals to be transmitted without interfering with each other. The 3.58 megahertz carrier is suppressed before it is added to the station's main carrier.

The luminance and chrominance signals contain all of the picture information. The color burst signal is added to the composite video signal on what is referred to as the *back porch* of the horizontal sync pulse. This is shown in Fig. 7-13. This signal does not contain any color information. It is used to synchronize the color oscillator in the receiver with the transmitted signal. The I and Q signals are transmitted by a double-sideband

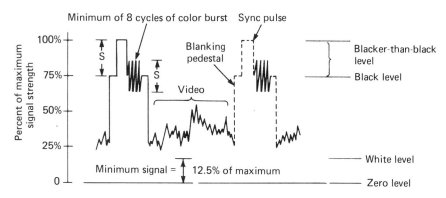

Figure 7-13. A line of the composite video signal. Included in this signal is the color burst, which is located on the *back porch* of the blanking pulse.

suppressed-subcarrier system. The carrier has to be re-created in the receiver before any detection action can occur. This is accomplished by use of a 3.58 megahertz oscillator in the receiver. In order to eliminate any distortion, it must be synchronized with the transmitted signal. The color burst signal does this job. These signals are added to the carrier in a manner similar to an FM-stereo station multiplex system. The name given this system is *colorplexing*.

7-6
Signal Paths

Once you understand what the electronic signal is and what the component parts of the signal are, you are almost ready to follow these signals through typical home-entertainment device block-diagrams. An intermediate step must be taken prior to looking at how signals are processed in these sets. This step involves looking at the various types of signal paths used in electronic devices. There are but a few of them. Knowledge of each of these types of paths and what happens to electronic signals is fundamental to a full understanding of how the radio, TV, or sound system works.

There are a few basic signal-path patterns used in most electronic equipment. The more complex sets use combinations of these patterns. Looking at inputs and outputs of specific blocks helps to identify which path pattern is being used. Each of these patterns is displayed in Fig. 7-14.

These six signal paths are not difficult to understand if you look at them as if the signal were entering the block from the left and leaving to the right. Most electronic schematics and block diagrams are drawn in this manner. There are arrowheads on each part of the diagram to help you understand how the signals move through the blocks. Let us now look at the six individual flow-path systems.

Linear. Linear means *in line*. In this system, there is only one path for the signal to take: each block must do its job properly if the signal is to be processed through the next block of the system. If any one block fails, then the system does not work properly. This flow path is the simplest to follow. The signal enters at the left side of the block diagram and leaves the first block from the right side. The signal progresses in this manner through each of the blocks in the set.

Convergent. When lines converge, they meet at some point. The same holds true for signal flow: two signals meet at a block in the set. Once inside this block, these signals may possibly be changed in size or shape. Our concern now is what happens to the signal flow from one block to another. At times there may be more than two signals converging at one block. When this occurs, it is still considered to be a converging signal-path system.

102

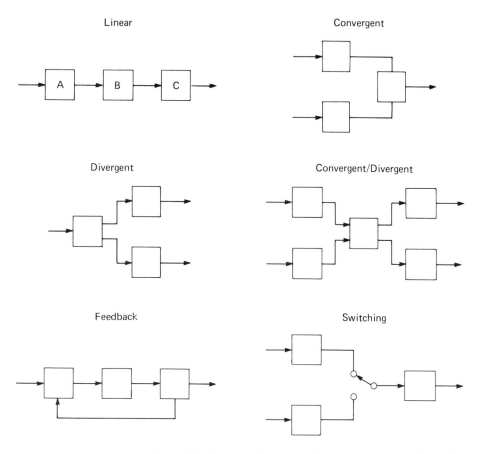

Figure 7-14. Signal-path systems found in electronic devices. The flow to the signal through each system is indicated by the arrowheads.

Divergent. In some cases, the signal flow is from one block into two or more blocks. This is called a divergent path system. Signals coming from a common block may be used to operate several other blocks in the set. An example of this is in the set's power supply. Operating voltages are established in the power supply block and sent to every other block in the set.

Convergent-divergent. At times several signals come together at one point, or block, in a set and are processed in that block. Then this modified signal is sent to several other blocks. This merging, processing, and separating is referred to as a convergent-divergent signal-path system. An example of this is seen in the block diagram of a TV set. The AM picture signal and the FM sound signal converge in the tuner section. They are processed as a common signal through the IF amplifier and the

103

detector, where they are separated and sent to specific blocks for further processing. In this example, more than one block is used for the converged signals. The procedure remains the same if a single block is used.

Feedback.   There are instances in electronic circuitry where a part of a signal is returned to a preceding block. This process is called feedback. Feedback is a system of sampling a part of the output and sending this sample back to another block in order to correct any errors or incorrect values in the set. One such circuit is the automatic gain control (AGC) system which tries to maintain a constant output from the set regardless of the strength of the received signal.

Switching.   A switch is a mechanical device used to select a specific path for a signal to use. There may be as few as two paths or there may be several paths to select from. The switch may be wired to select one path from several. An example of this is the input selector of a stereo system. Here, the input selector switch selects the desired input, either AM, FM, tape, or phono. In other situations, a switch may be used to direct the signal from a block to one of several different blocks. One such case is the switch which selects either loudspeaker or headphones in a radio output circuit.

Each of these six signal path systems are in general use in home-entertainment equipment. The qualified technician becomes familiar with basic set block-diagrams and is able to relate these blocks with how the electronic signals move through the set. This knowledge helps greatly when he has to service a set. He must have the ability to determine signal flow through block diagram analysis, and he must have some knowledge of how to troubleshoot each type of flow pattern. Learn the signal flow-paths now. How to use this information for basic troubleshooting is covered in a later chapter.

SUMMARY   Electronic signals are used to carry information from a sender to a receiver. The signal may be a very simple pulse of electrical energy or it may be a very complex signal consisting of a carrier modulated by several different subcarriers. Signals are processed by the device which uses them. The signal may appear in a different form at the output of the set. The whole purpose of the signal is to get a *message* to the receiver in a form that is processed, and is ultimately seen or heard by the person receiving it.

A radio or television signal is made up of a continuous signal called a carrier and the information contained in the signal. The information is added to the carrier by means of mixing the two signals. This process is called modulation. There are two basic types of modulating systems. These are amplitude modulation and frequency modulation. In an amplitude modulation system, the modulating signal causes the amplitude of the carrier to vary as the shape of the modulation varies. Frequency modulation produces slight variations in the frequency of the carrier wave and does not vary the amplitude of the carrier in any manner.

FM stereo or television signals have more information on them. These signals use a variety of modulating systems in order to transmit all of the information required to reproduce the information in the receiver. Methods of sending modulation on one sideband of a carrier have been developed and are in use. These methods include double-sideband suppressed carrier, single-sideband suppressed carrier, and vestigial sideband transmission. FM stereo uses double-sideband suppressed carrier to carry the stereo portion of the signal on the main carrier. TV broadcasts use vestigial sideband transmission for the picture carrier. FM transmission is used for the sound portion of the TV channel information. The method of adding a modulated subcarrier to the main carrier is called multiplex modulation. This is often used with complex signals. Both FM stereo and color TV signals are multiplexed signals.

QUESTIONS

1. Define the term *electrical signal*.

2. Define the term *signal path*.

3. List each of the six signal-path systems and explain how each works.

4. State five characteristics of electrical signals.

5. What electronic tester is best used to measure the characteristics identified in Ques. 4?

6. How are complex waves formed?

7. What is the multiplex form of modulation?

8. Name each method of modulation presented in this chapter. Briefly explain what waveforms for each system would look like.

9. What purpose does the sync pulse serve in a TV set?

10. What is the purpose of the color burst waveform and where is it located on the composite video signal?

Signal Paths: Audio Devices

8

Success in the home-entertainment repair field is based upon the ability of the repair technician to quickly fix a nonoperating unit. In order to be successful, this person must first have a strong background of knowledge in how the unit is supposed to function. Secondly he must be familiar with the electronic theory involved. Third, he must know how to select and use electronic testing and measuring equipment in order to diagnose trouble areas. Let's take a closer look at each of these three areas of knowledge.

The first area identified is knowing how the unit is supposed to function. The beginning of this book is devoted to the study of functional block diagrams of typical home-entertainment devices. Each block in the unit is named, and the function of the block is described. Much emphasis is placed on the learning of each block diagram. The author also identifies similarities among related blocks in the various home-entertainment units presented.

Part of the block-diagram analysis of functioning units is the study of how electronic signals are processed in each of the typical units. Here, as with the functional block diagram, the study is based upon the input and output signal waveforms in each block. The signal being studied is an electrical wave that carries a form of intelligence. This intelligence may be the audio portion of AM radio. It may also be the audio portion of FM radio or of television. Also classified as intelligence is the picture information in the TV set as well as the synchronizing pulses. In other words, the intelligence is all of the information delivered by the carrier from the transmitter to the receiver. Once delivery has been accomplished, the carrier is eventually discarded, and the information delivered to the receiver is used to reconstruct the original message. A successful repair technician requires a good working knowledge of what the incoming signals are made up of and what these signals look like on the basic testing equipment used to display them. This same technician is required to know how each signal is processed through the set. He must be aware of the frequencies used for each signal. He has to be able to read the literature and use the testing equipment in order to trace the signal through a specific set.

The second area in which knowledge is required for success in the home-entertainment repair field is how the electronics in the set produces the amplification of the signal or any of the other modifications of the signal that occur. This knowledge is based upon a good basic knowledge of electronic theory. The technician has to know the purposes of the individual components used in the production of the set. He should be knowledgeable in how the component parts work.

Some of these parts are *active* devices, that is, they change during their period of operation. Typical active devices found in home-entertainment devices are transistors, integrated circuits, and vacuum tubes. A solid background in electronic theory as it applies to these active devices is required if the technician is to be a repairperson instead of someone capable of nothing more than changing a module or board.

Other parts are considered to be *passive* devices. They do not change during operation. The operation of these devices also has to be understood. Passive devices are as important to circuit function as the active devices are. The study of these components is covered later in this book.

The third area of knowledge required for efficient and successful repair is of how and when to use electronic testing and measuring equipment. This is a difficult topic to present in a book because so much of the ability to use test and measuring equipment is based upon practice with the equipment. Theory in this area is useful, up to a point. Books can present material on how and why to use specific equipment. They can also discuss the theoretical makeup of specific units. However, learning to *use* test equipment can be compared to learning to drive a car. Many a person is able to grasp the theories of automobile driving, but it isn't until one is behind the wheel and actually driving that the application of these theories has any real meaning. This same holds true for using test equipment. Practice and more practice is the only way to develop skills with test equipment. The author suggests combining laboratory, book, and class study to help improve skills in the area of test equipment utilization.

Words can only go so far in teaching electronic repairs. The purpose of this book is to provide introductory material so that the student is able to start to apply this knowledge to the occupation of home-entertainment electronic repair. Much time is spent in developing concepts on how things work as well as why they work in the manner in which they do. This section of the book covers signal paths. It tells how the signal is processed through the set by looking at inputs and outputs of each functional block in the set. Later chapters discuss active- and passive-component theory. Still later chapters tell how the electronic components function inside each block. Let's start our consideration of signal path identification by looking at some of the terminology associated with electronic signal amplification and at typical signal voltage levels produced by input transducers used with audio amplifiers.

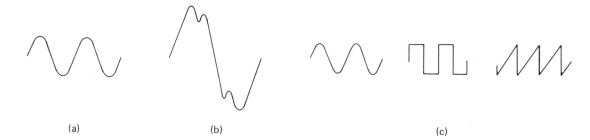

(a)   (b)   (c)

Figure 8-1. Signals used in audio devices: (a) sine wave; (b) sine wave with harmonics; (c) the sine wave, square wave, and triangular wave, as produced by an electronic signal generator.

Signals commonly used in audio amplifiers are based upon a sine-wave shape. These are displayed in Fig. 8-1. The simplest signal would appear as a pure sine wave at the frequency of the tone, as displayed in Fig. 8-1(a). More complex wave forms are made up of fundamental frequencies and the *harmonics*, or multiples, of these frequencies. A wave similar to this is shown in Figure 8-1(b). Test equipment used to produce electrical signals for testing amplifying devices may produce up to three basic waveforms. These are the *sine* wave, the *square* wave and the *triangular* wave. These are shown in Fig. 8-1(c). An often used frequency for these waveforms is 400 hertz. Some signal generators have the capability of producing only this frequency and are not capable of producing the entire range of audio frequencies of 10 hertz and 20 kilohertz. The selection and application of the proper signal and frequency depends upon the type of test being conducted. A simple test to determine if the amplifier is able to amplify the test signal would most likely use a 400 hertz test signal. If the test was to determine the frequency response of the amplifier, then a full-range variable-frequency generator would be required. Here, the technician must know how to select the proper test equipment for the job. This information is often provided in the manufacturer's service literature. For the purposes of this book, the author will use a simple sine wave in order to indicate an audio frequency signal.

8-1
Signal Levels
Specified in the literature published relating to input transducers utilized in home-entertainment equipment is among other things, the amount of output voltage developed by these transducers. This output voltage may be classified in two ways. One method is based simply on output voltage level. Phonograph cartridges are rated in this manner. The range of output voltages available is wide, and depends upon the physical makeup of the cartridge. Some units have a very low output voltage. Often this voltage is about 0.009 volt. Other phono cartridges have output voltage levels as high as 5 volts. There are still other units whose output voltage levels range between these two extremes. Each phono cartridge produced has its own set of specifications for output voltage and frequency response. As a general rule, it is safe to say that high quality reproduction from phonograph records is developed with a low output-voltage magnetic cartridge. The higher output-voltage cartridges do not normally produce better fidelity sound. Use of low output-voltage cartridges does require more stages of amplification in order to produce large amounts of sound in the loudspeaker. Tape recorder heads have even lower output-voltage levels than do phono cartridges. A normal output voltage for a tape head is in the area of 0.001 volt.

Microphones use the second method of rating output-voltage levels. This method is related to the amount of sound the microphone is able to pick up and convert into electrical energy. This rating is in decibels. The *decibel*, as used in electronic work, represents a ratio of power, voltage, or current. This ratio can be either a positive or a negative value. Often the gain or amount of amplification of a set or device is expressed in decibels.

8-2
Electric Power
Electric power is defined as the time-rate of doing electrical work (or the time-rate of consuming energy). In earlier chapters the terms *voltage* and *current* were introduced. Voltage is identified as a force that is able to make electrons move. Current is defined as the quantity of electrons moving past a given point in a standard time-reference of one second. These two factors of voltage and current are directly related to electrical power. *Power* in electrical work is determined by multiplying the voltage by the current. The formula is power = voltage × current. Abbreviations used for these terms are P for power, E for voltage, and I for current. Substituting these letters in the power formula gives us $P = E \times I$. The unit of electrical power is the watt. There are ways of determining either power, voltage, or current. If two of the three values are known, then the other one may be easily computed. Figure 8-2 shows what is called the *power triangle*, which shows the relationship of the three components used to figure power. It is used in this manner: if you wish to find one of the three factors and you know the other two, then place your finger over the unknown and the math formula is there to use. For example, if you know the voltage and current, then placing your finger over the P on the triangle shows the $P = I \times E$. To find the current when the power and

voltage are known, place your finger on the *I*. The formula now is *I* = *P/E*. This method is used to find any of the three quantities relating to electrical power, and is a simple way to find the unknown factor. Electrical power is important in all areas of electronics.

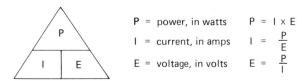

P = power, in watts   P = I × E

I = current, in amps   I = $\frac{P}{E}$

E = voltage, in volts   E = $\frac{P}{I}$

Figure 8-2. The power triangle shows the relationship among power, voltage, and current.

An example of the use of this formula and the power triangle would be in finding the current flow in a TV set rated at 650 watts. The voltage available from the power company is 120 volts. Two of the three factors in the power triangle are now known. The unknown is the current. Placing a finger over the *I* will reveal the formula *I* = *P/E*. Substituting numbers for the two known values gives I = 650/120. The answer to this problem is a little more than 5.4 amperes of current.

8-3
Gain

The term *gain* as used in electronics refers to a relationship between the input and the output. One may refer to the overall gain in an audio amplifier. Gain may also be related to a specific transistor or vacuum tube stage in the amplifier. There are three kinds of gains that may be determined. These are voltage gain, current gain, and power gain. Regardless of the specific kind of gain involved, the method of finding the amount of gain is the same. Simply divide the output value by the input value. Look at Fig. 8-3 as a reference. This figure shows the various voltage levels in a tape recorder amplifier. Using the formula, Gain = Output/Input, find the gain for the first block. In this case, as with the power formula, substitute numbers for the known values. An important thing to remember is that one has to stay with the same kinds of numbers. Do not use millivolts and volts. All figures must be in the same set of values. If the input was 10 millivolts, change it to be .010 volts, so that all figures are in the same type of reference. To determine the gain of the first stage,

$$\text{Gain} = \frac{\text{Output}}{\text{Input}}$$

Substituting values for the output and the input gives us

$$\text{Gain} = \frac{0.042}{0.002}$$

110

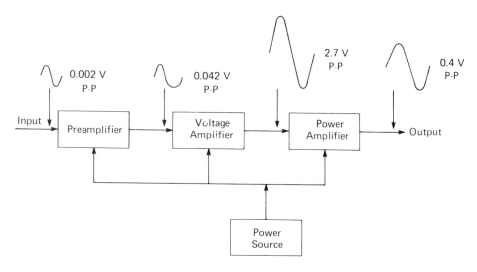

Figure 8-3. Signal voltage values in a typical amplifying system. The signal is amplified from the input valve of 0.002 volts to a value of 0.4 volts at the final stage.

Solving this tells us that the gain for this stage is 21. Now repeat this for each of the other blocks. Gain for each of the blocks is 21, 64, and 0.15.

These figures tell the technician the amount of signal amplification that occurs in each block of the unit. Using an oscilloscope, the technician is able to measure the input and output waveforms of each block and compare these to the manufacturer's specifications. When a block has less than the specified amount of gain, then the technician starts to look for some trouble in that block.

There are three methods of finding total gain in a unit. One way is to measure the input signal and compare it to the output signal. A second way is to find the product of the gain of each stage. To find the product of the gain of each stage, multiply the numerical values of gain for each stage together. Using the block diagram in figure 8-3, this would be done by multiplying 21 by 64 by 0.15. The result should be the total gain of the unit, or 202.

This value of 202 represents the total signal voltage gain of our model amplifier. This figure includes the gain of the power amplifier stage. The power amplifier stage normally does not have any voltage gain. It is used to produce a large amount of signal current rather than the signal voltage. The signal current is required in order to operate the loudspeaker. As a result, the signal voltage gain of the stages or blocks prior to the power amplifier is large. This results in a total gain value of almost 1,350 for these stages. When the voltage gain for the final stage is included, then the overall gain of the amplifier is reduced to a value of 202 ($21 \times 64 \times 0.15 = 202$). A person who is not familiar with how the amplifier operates might assume that there was something wrong in the system because of the low overall-gain figure. The technician, knowing how the system is supposed to work, would accept these values as being representative of a properly functioning amplifier.

The third method of finding total gain is to convert the power gain figures into decibels and to simply add up the amount of decibel (abbreviated *dB*) gain for each stage.

Refer to Fig. 8-4. This chart shows how to convert power gain figures into decibels. In any case, if the output values are equal to the input values, then a ratio of 1:1 is established and there is no gain. If, however, the input power is greater than the output power, then there is a power loss and the result is a decibel gain figure of less than one, or a negative value. Taking a practical example, an input power value of 10 and an output power value of 1 results in a power gain of 0.1. Using the chart, this converts into a gain figure of −10 decibels. At the other end of the line, if the input value were 2 and the output value 200, the result would be a gain figure of 100. This converts into a power gain of 20 decibels. All three

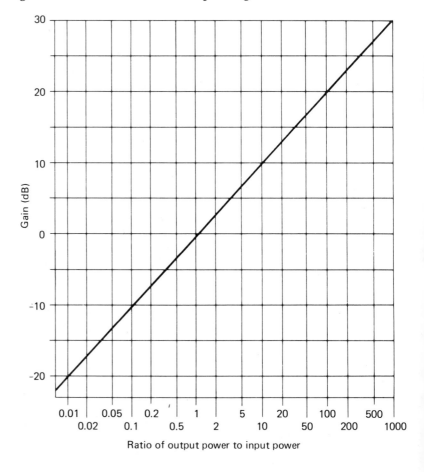

Figure 8-4. A chart used to convert power gain into decibels.

of these methods are in use today when determining gain. However, most literature concerned with how to repair home-entertainment devices does not use decibel ratings when showing individual-stage gains, but offers both input and output waveform values for the technician to use in his repair work.

Since the power gain values are commonly used, let's see how they apply to our model amplifier. In order to do this, we will have to add some more information to our system, that is, the amount of power that is used in each block of the amplifier. For this, we will establish some typical values. These are shown in Fig. 8-5. First, convert the power ratios into number values. This is done by dividing the output power by the input power (Power gain = Output power/Input power). Do this for each stage. The results are values of 10, for the first stage, 20, for the second stage, and 1, for the third stage. These are then converted into decibel values by using the chart in Fig. 8-4. A power gain of 10 is equal to 10 decibels. A power gain of 20 is equal to about 12.5 decibels and a gain of 1 is equal to 1 decibel. These values may be added in order to determine the total power gain of the set (10 dB + 12.5 dB + 1 dB = 23.5 dB). Converting this back into a power value shows that 23.5 decibels is equal to a power gain of about 200. This figure could be cross-checked by multiplying the gains of each stage (10 × 20 × 1 = 200). In this case, the answer was reasonably easy to determine. In amplifiers with many stages, it is often easier to add the decibel gain for each stage than to multiply the gains for all the stages.

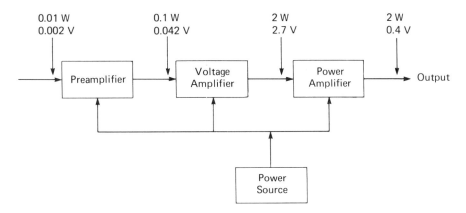

Figure 8-5. Signal voltage values and power values for a typical audio amplifier.

When analyzing these illustrations, it becomes obvious that almost all of the voltage gain of the signal in an amplifier occurs in the voltage-amplifier blocks. The power-amplifier block is used to increase the total signal power in order to operate the output transducer, or loudspeaker. Amplifiers with larger power ratings may require additional stages of voltage amplification in order to develop a signal strong enough to operate the power-amplifier block.

8-4
Amplifier Input
Requirements

Any amplifier is only capable of increasing the signal to some predetermined level. Each amplifier has specifications, which are developed by the design engineers. Both input and output levels are spelled out in the performance specifications. Each amplifier has an input signal-level value. What this means is that when the input signal is equal to the design value, then the amplifier is able to produce the specified amount of signal power at the output terminals. If the input signal is less than the design value for the amplifier, then the output to the speakers will be less. On the other hand, if the input signal is too large, then the amplifier will not be able to reproduce all of the signal. The result of too large a signal is often distortion. A distorted sound is not a faithful reproduction of the input signal form, and is an undesired effect.

When a multiple-input sound system is used, each input signal has to be matched to the input of the amplifier. For example, a home-entertainment system may have several input systems. They may include a record player, a tape player, and an AM-FM tuner. Each of these devices may have a different output level. In order to bring all of these devices up to a certain signal level, a device called a preamplifier is used. The preamplifier is really a voltage amplifier that is designed to take a very low level signal and to amplify this signal so that it may be used by the amplifier. A system using multiple inputs and some preamplifiers is shown in Fig. 8-6.

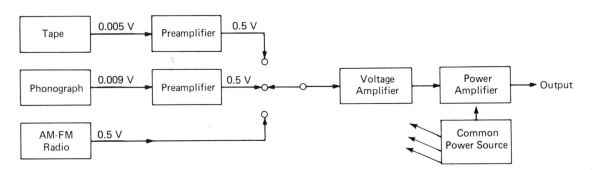

Figure 8-6. Block diagram of a multi-input audio system. Note the use of preamplifiers to increase the signal level to a common value.

This system uses two preamplifiers. One is used with each low signal-level output device. Typical output levels for tape recorders are on the order of 0.005 volt. Record player cartridges may have output levels of 0.009 volt. Tuners often have outputs of 0.5 volt. These signal levels have to be amplified to some common level in order for the amplifier to produce its *rated* power output. This is accomplished by inserting a signal preamplifier between each of the lower output devices and the amplifier. Each preamplifier is adjusted to match the signal level of the radio. In this manner the output power of the amplifier is always the same even though each input unit produces a different signal output level.

8-5
Power and
Power Transfer

Electrical power is directly related to the ability to perform electrical work. You have seen illustrations showing how power is used in each block of an audio amplifier system. Keep in mind that this power is not moving from one stage to another stage in the amplifier. It is used in a different manner. By definition, an amplifier is a device that uses a small amount of power in order to control a larger amount of power. In other words, the input power to each stage in the amplifier is less than the output power. The input power is used to *control* the output power. As the input power increases, so does the output power. If the input power decreases, the output power also decreases. This power has to be transferred from one stage onto the next stage of the device. The output of one stage of the amplifier must be matched to the input of the next stage for efficiency purposes.

In *mechanical* work, if a mismatch occurs, there is a power loss that appears as heat somewhere in the system. It may be in a bearing or in the motor driving the unit. This same type of energy waste occurs in *electronic* work. A mismatched system has a lower output than a system in which all stages are matched to each other. This problem is one that is normally handled by the engineers who design the sets. However, the repair technician must be aware of this because he, too, has to match systems. The most common place that this occurs is in the output circuits of amplifiers. Loudspeakers must be connected and matched to amplifiers in order to efficiently transfer the signal. If this is done incorrectly, the resulting mismatch often produces heat in the output stage and a lower signal level in the speaker. (See Fig. 8-7.) The formula for transfer of power from one stage to another is:

$$\text{Power out} = \text{Power in}$$

Both input power and output power have to include any and all energy. The signal represents energy. The heat due to mismatch also represents energy. It is best to attempt to have maximum signal and minimum heat in both the input circuit and the output circuit if maximum signal is to be transferred from one stage to another stage. As the heat increases in value, the amount of signal is reduced. A power value of 10 watts could be made up of 8 watts of heat and 2 watts of signal. This would produce a very low level of output signal. On the other hand, if the heat was only ½ watt and the signal 9½ watts, then there would be a good transfer and a high level of signal produced in the next stage.

The power formula states that power equals voltage times current. If

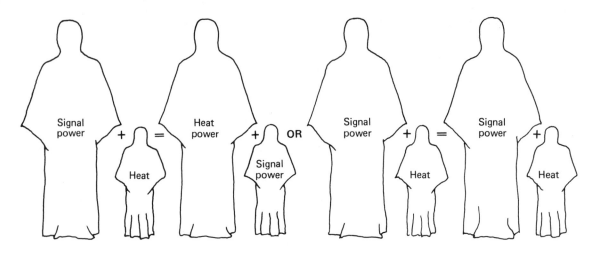

Figure 8-7. Maximum signal transfer occurs when the signal is large compared to the heat (wasted energy) produced in the set.

the voltage rises, then a smaller amount of current is required to produce the same amount of power. For example, 10 watts of power are produced by 5 volts and 2 amperes. Ten watts of power are also produced by 2 volts and 5 amperes. Six volts and about 1.6 amperes also produces 10 watts. In other words, the actual values required to produce 10 watts of power can vary, but the power produced will be the same. This is exactly what happens in the power output stage of an amplifier. The input signal has a high voltage level and a low current level. The signal is transformed through a matching device (tube, transistor, or transformer) to the output device. In most instances, the voltage level is reduced quite appreciably while the current level increases greatly in order to operate the output device.

Another way of stating this is that maximum power transfer occurs when the load (or output) resistance is equal to the resistance of the source. This statement holds true in all kinds of circuits. It also is true for every kind of power source. The most convenient way to describe this action is by use of a direct-current source and a resistive load. The power in the circuit depends upon the applied voltage and the amount of current in the circuit. The current, in turn, depends upon the applied voltage and the total resistance in the circuit. If there is very little internal resistance in the power source, then the resistance of the load determines how much current is flowing in the circuit. As the total amount of resistance increases, the total current flow in the circuit decreases. The result of this action is less power developed in the output, or load circuit. This is very true when a battery power source is in use. Batteries tend to develop high internal resistances as their internal parts age and deteriorate. This is illustrated in Fig. 8-8.

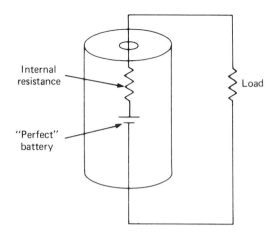

Figure 8-8. Maximum power transfer occurs when the internal resistance of the battery is low.

The power source consists of the battery and the internal resistance of the battery. As the internal resistance of the battery increases, total resistance of the complete circuit consisting of the battery, the internal resistance, and the load increases. The result is less current flow as well as less voltage available at the load. This is a classic example of how batteries tend to fail, of what causes lamps in flashlights to grow dim, and why battery operated radios tend to lose their loudness.

8-6
Impedance

In the discussion about power transfer, a direct-current source is used to show how power transfer is accomplished. Life is not always this simple. In many circuits found in home-entertainment devices, the kind of transfer that occurs is that of the electronic signal. The very nature of the waveform used to indicate a signal shows that the signal voltage is constantly changing. This constant change affects the circuit by introducing a factor called *reactance*. Reactance is the opposition to electron movement in an AC circuit. The reactance in the circuit due to the constantly changing values combines with the resistance in the circuit to produce another value. This new value is called *impedance*. Impedance is the total opposition to changes in circuit values due to reactance and resistance. The maximum power-transfer rule also applies to a changing circuit. All that is required is that one modify the statement so that it reads: maximum power transfer occurs in a load when its impedance is equal to the impedance of the source. The most common use for this rule is in the connecting of loudspeakers to an amplifier. Home-entertainment amplifiers have output terminals connected to either a 4-, 8-, or 16-ohm impedance source. The speakers are also rated in this manner; most quality units are rated at either 8 or 16 ohms. Output terminals on the speaker should be connected to equal value terminals on the amplifier in order to properly match the source to the load. Attempting to operate the amplifier without any speakers or load could produce a large amount of current and heat in the output stage. This, in turn, could destroy the output transistors and other components used in the output stage. For this reason, it is wise to be certain that the speakers are connected to the amplifier before power is applied to it.

SUMMARY

The electronic waves which carry intelligence, or a message, are called signal waves. These waves are produced by either a transmitter, a microphone, or a TV camera. Audio amplifiers use signals normally produced by a microphone, a phono cartridge, a prerecorded tape, or an AM-FM tuner. Electronic signals may appear as almost any shape or form. For the purposes of this chapter they are limited to a sine waveform. Normally these waves are much more complex.

Signal levels will vary with the design of the transducer. The range of the signals may be as low as 0.001 volt for a tape head or as high as 0.5 volt for a phonograph cartridge.

Electric power is defined as the time-rate of doing electrical work. The unit of electrical power is the watt. The power formula is: Power = Voltage × current. This formula is used to determine the electrical power used by all electrical devices.

Gain is the relationship of the output to the input. The formula for gain is: Gain = Output/input. When speaking in terms of gain, it is necessary to define the reference either as voltage gain, power gain, or current gain. It is also necessary to identify the kind of device used—this could be a single stage or it could be a complete unit, such as an amplifier or a radio. Gain may be determined using any of three methods. One is to multiply end-stage gain by every other stage gain figure. Another way is to convert the gain for each stage into decibels and then add up the figures. A third method is to compare input-signal measurement to output-signal measurement. All three methods are in use in electronics.

Some audio amplifiers are multi-input devices. When this is the case, it becomes necessary to use preamplifiers in order to present the same signal voltage level, from all inputs, to the amplifier. The preamplifier is connected between the tape player or phono and the amplifier. It is adjusted in order to present the same amplitude signal to the input of the amplifier.

The purpose of an amplifier is to increase the size of the electrical signal in order to reproduce the signal at the loudspeaker. Initial stages in the amplifier may increase the amplitude of the signal to several hundred times its input value. The output stage of the amplifier is designed to produce power in order to operate the loudspeakers.

Power must be transferred from one stage to the next stage in any electronic device. A simple rule states that the power going from the stage must equal the power entering the next stage. If these stages are not properly matched, then some of the power is wasted as heat. The ideal transfer situation minimizes the heat loss in order to maximize the amount of power being transferred.

Impedance is a combination of circuit resistance and circuit reactance. This forms the total opposition to changes in circuit values when the values are constantly changing. A circuit that is processing an electrical signal has constantly changing values itself. As a result, it is the impedance of the circuit that needs to be considered, rather than the resistance, when power is transferred.

QUESTIONS

1. State the three major areas of knowledge required for success in the electronics repair field.

2. Name three basic waveforms used in electronics.

3. State the typical signal levels for:
   (a) a tape head
   (b) a magnetic phono cartridge
   (c) a crystal phono cartridge

4. State the power formula.

5. Find the power used in a TV set operating from a 12 volt source and using 4 amperes of current.

6. Find the amperes used in a TV set operating from a 120 volt source and rated at 600 watts.

7. The input signal of a stage in an amplifier is 0.003 volt. The output signal is 1.2 volts. How much voltage gain occurs?

8. The input power of a stage in an amplifier is 0.005 watt. The output power is 0.050 watt. What is the power gain?

9. Using the values given in Ques. 8, determine the gain in decibels for this stage.

10. State the rule for power transfer.

11. What is the purpose of a preamplifier?

12. Why is it important to use an impedance matching device between stages of electronic units?

Signal
Paths:
Radio
Receivers

The introduction of the radio as a home-entertainment and communication device has done much to bring the various sections of the world closer together. Consider for a moment the length of time that used to be required to send a message by horse and rider, or boat, from one part of the country, or world, to another part. In some cases, many weeks elapsed before the message was delivered. Today we can have almost immediate communication. Radio and electrical signals travel at the speed of light, 186,000 miles per second. Newsworthy events anywhere in the world are reported by radio almost instantly after they occur. The introduction of the radio transmitter and receiver have made this possible. Refinement of these radios from units requiring house current or heavy batteries to units which are very small and lightweight have helped make us a communications-conscious world.

One of the earliest types of radio communication used the Morse code. A radio broadcast transmitter was turned on and off in step with the dots and dashes used as code symbols. The person listening to the code symbols had to translate the symbols into letters. The letters were then grouped into words. This type of communication could not reach the person who did not know, or care to learn, the code system.

Further refinements and experiments with radio transmission were made. From those experiments came a system which was adopted for mass communication. That system took the audio signal carrying the intelligence and blended it with the radio-transmitter carrier signal. This blending process is called modulation. There are several kinds of modulation systems. These were discussed in Chap. 7. Figure 9-1 shows one basic type used in entertainment radio receivers. The modulating information changes the form of the carrier signal. Part (a) of this figure shows the radio-station carrier signal. It is composed of a sine wave at the station's assigned frequency. Broadcast station carrier signals are sine waveforms regardless of the station's operating frequency. Part (b) of Fig. 9-1 shows the modulating waveform. It, too, is a sine wave. However, it is of a much lower frequency. The result of the modulating process is shown in Fig. 9-1 (c). The modulating waveform causes the *amplitude* of the carrier to vary. The carrier's amplitude takes on the shape of the modulating signal. This type of system is called *amplitude modulation*. Amplitude modulation is used with several different types of transmitting systems. One of these is referred to as *AM radio*. Home-entertainment AM radio stations have been assigned a range of operating frequencies by the Federal Communications Commission. These range from 540 kilohertz to 1.6 megahertz. This spectrum of frequencies relates to the assignment of the carrier frequency. In other words, each AM radio station is assigned an operating frequency for its carrier somewhere within the AM broadcast band. Efforts are made to insure that no two stations will interfere with each other's broadcasting.

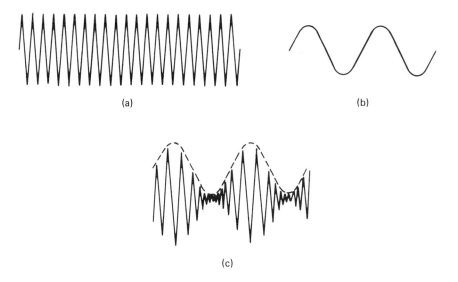

(a)

(b)

(c)

Figure 9-1. The carrier signal (a) and the modulating signal (b) combine to form an amplitude modulated carrier (c).

9-1
Signal Processing

Refer to the block diagram in Fig. 9-2. This is the same AM radio block-diagram shown in an earlier chapter. Most AM radios marketed today use circuitry which form these blocks. This type of receiver uses the

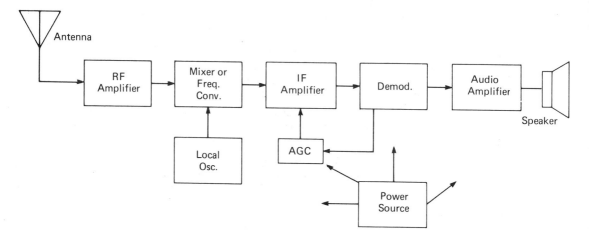

Figure 9-2. Block diagram of a superheterodyne radio receiver.

heterodyne principle. Heterodyning is a process in which two signals are mixed in order to produce another signal. This differs from a modulation system in that the resulting wave in a heterodyne system keeps the modulation. This system produces a carrier at a new frequency without changing the information being transmitted. This type of radio is called a superheterodyne radio, or just *superhet* radio. The easiest way of understanding how this system works is to look at the section of the radio in which all of the tuning and station selection occurs. The tuner consists of three blocks: the RF amplifier, the mixer, and the local oscillator. The purpose of these blocks is to receive the signal being broadcast, heterodyne this signal with one produced in the local oscillator, and produce a third signal called an intermediate frequency (IF) signal. This IF signal is then processed through the radio and eventually the original information as it was broadcast, is heard in the speaker.

Figure 9-3 shows how two carrier signals are mixed in the superhet system. Both F_1 and F_2 are to be heterodyned in the mixer block of the receiver. The result of this action produces four signals. These are: F_1 at 1.5 Megahertz, F_2 at 1 megahertz, the sum of these two frequencies, or 2.5 megahertz, and the difference between the two frequencies, or 0.5 megahertz. If the incoming signal at F_2 was modulated, then both the sum and difference signals would contain modulation. The modulated carrier signal received at the antenna is changed into an intermediate frequency (IF) signal in this manner.

Look at Fig. 9-4. This is a typical AM receiver block-diagram. Signal waveform information has been added to this diagram in order to show how the received signal is processed in the radio.

Antenna. The antenna of any receiver is not a true block. It is used as the input transducer. Electrical waves from the transmitter are broadcast

122

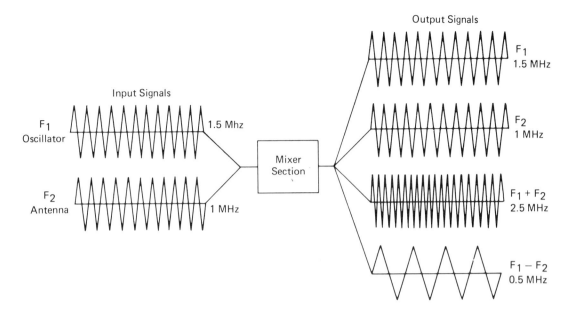

Figure 9-3. The heterodyning process produces four signals through the action of the mixer stage.

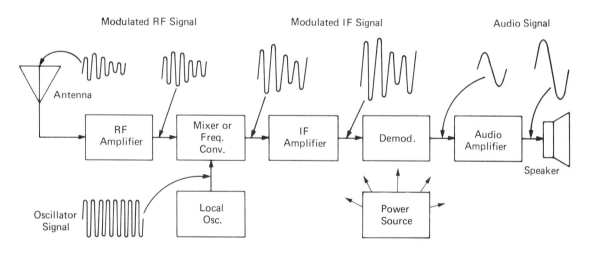

Figure 9-4. Signal processing in an AM radio.

into the air. They are electromagnetic waves. As these waves strike the antenna, a small electrical signal is induced by the waves onto the antenna. This signal may be as small as 0.00015 volt. It may also be larger, depending upon the distance between the transmitter and the receiver and the power of the transmitter. The ability of any receiver to *pick up* an electromagnetic signal from the air is called the receiver's *input sensitivity*.

As a rule, better quality receivers have lower input sensitivity ratings than do the poorer quality receivers.

Radio frequency amplifier-selector. The signal received from the transmitter consists of the carrier wave, on its assigned frequency, and the information being sent from the station. This wave is in the form of an amplitude modulated wave. The shape of the wave depends upon the shape of the modulating signal. The overall amplitude of the signal depends upon the amount of signal being received.

The first block of the radio is used to select the proper broadcast station signal from the many stations on the air. This is accomplished by use of a *tuned circuit* in the radio. This circuit is able to select one specific frequency and reject all others. The circuit has the ability to change the selection frequency in order to move to a different station. The tuning dial of the radio is attached to the circuit. As the dial is turned the selection frequency of the radio changes.

Some radios use an amplifier in this block. When the signal is received and selected, it is also amplified. Other radios, particularly those sold at low prices, do not use an RF amplifier. Because of this omission, the ability of the radio to receive distant stations is very limited. Note on the block diagram of the radio that it uses an RF amplifier. The received signal is amplified as it is processed through the radio. The output of the RF amplifier-selector is sent to the mixer block.

Local oscillator. The local oscillator of the receiver generates its own carrier signal. No modulation is found on this signal. The local oscillator is designed to be on when the radio is on. Its frequency is tunable. Radios using the local oscillator have the tuning control for both the local oscillator and RF amplifier-selector on one common knob. Both blocks are tuned at the same time. The output signal from the local oscillator is also sent to the mixer block.

Mixer. The received signal and the local oscillator signal are heterodyned in this block. The result of this action produces another signal. This was called the *difference* signal earlier in this chapter. Technically, it is called an intermediate frequency, or IF, signal. This is because its frequency falls between the carrier frequency being received and the radio frequencies. Most AM radios produced in the United States use an intermediate frequency of 455 kilohertz. The only exception to this is that a great many car radios use an IF of 262 kilohertz.

Tuning both the RF amplifier-selector and the local oscillator at the same time and rate produces the fixed IF. This same principle is applied to other kinds of receivers used for home entertainment. These will be discussed later. Another common name for this block is the *converter*. This name is used because the block accepts two signals and converts these signals into another usable signal at the intermediate frequency.

IF amplifier. This block amplifies the IF signal and its modulation. More amplification occurs in this block than in any other single block in the radio. The reason for this is that it's much easier to amplify a signal through a stage in which there is no variable tuning occurring than through a stage with variable tuning, such as the RF amplifier. There are several tuned circuits in the IF amplifier block. These are permanently tuned at the IF frequency of 455 kilohertz.

Detector. The detector block separates the audio signal from its carrier. The shape of the modulated signal was determined by the shape of the audio information contained in the signal. In its simplest form, the detector cuts the signal in half horizontally. It also removes the carrier signal, leaving only the audio portion of the wave. This action is illustrated in Fig. 9-5. Half of the modulated signal is permitted to pass through the detector. The remaining signal has the shape of the modulation. Another circuit filters out the IF carrier, leaving only the audio.

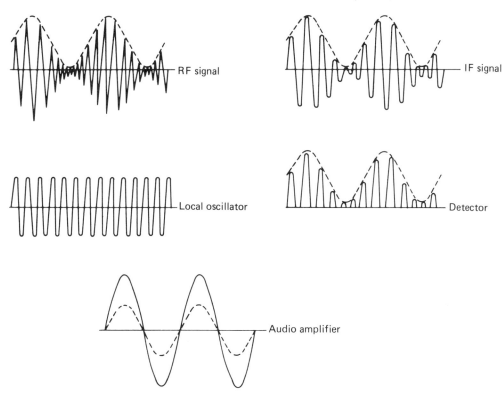

Figure 9-5. Signal processing showing how the modulation information is recovered in the receiver.

Audio amplifier. Little more needs to be said about the audio amplifier block. Its performance is identical to that of other audio amplifiers described in the previous chapter. Look again at Fig. 9-4. The small audio signal is amplified and then, through an impedance matching device, the signal is presented to the speaker. In the speaker, the electrical signal is changed into sound waves.

Stage gain. Figure 9-6 shows the kinds of gain possible in various blocks of the AM radio. These figures refer only to voltage gain. Reference to power gain is omitted. Note that the mixer has 2½ times the gain found in an RF amplifier. The IF stage has 8½ times the gain of the first stage. The audio block has almost as much gain as the IF amplifier block, while the audio output block has a lower value of voltage gain. The result of all of this amplification is an amplification factor of 25,500,000 times the size of the input signal. Looking at the radio sections which contain a modulated carrier, we see that most of the amplification or gain is occurring in the IF amplifier block. Other factors, such as the amount of signal received at the antenna, and the position of the volume control in the radio, may reduce this gain figure considerably. The illustration shows how and where the gain occurs in a typical AM radio receiver.

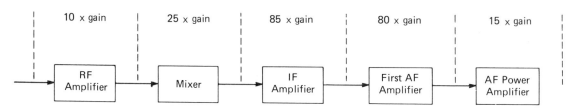

Figure 9-6. Typical voltage gains in specific blocks of a super-heterodyne radio.

9-2
The FM Radio

Perhaps the most convenient way to start a discussion of the FM radio signal is to look at the difference between AM and FM signals. AM, as you recall, stands for amplitude modulation. In this system the carrier takes on the shape of the audio signal. This is shown in Fig. 9-7. Part (a) of this illustration shows the station carrier waveform. Part (b) displays the audio tone which modulates the carrier signal. If the carrier is amplitude modulated, then a wave like that shown in part (c) is formed. However, if the audio signal modulates the *frequency* instead of the amplitude, then a wave like that displayed in part (d) is formed. In a frequency-modulated carrier system, the rate of the carrier signal is changed by the modulating signal. This modulation is a very low frequency change. The actual shape of the altered wave is dependent upon the exact signal modulating the wave. This system was developed in order to overcome the problems of low quality and possibly high atmospheric noise which are part of AM reception. Static, as received on a radio, results from impulses of electrical signals. These amplitude modulate the carrier. They are amplified in the

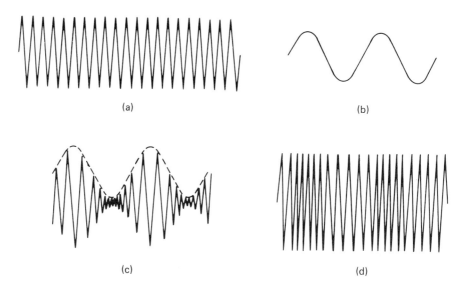

Figure 9-7. Modulation: (a) RF carrier; (b) AF signal to be transmitted; (c) amplitude-modulated carrier; (d) frequency-modulated carrier.

receiver, detected, and processed through the audio blocks. They turn up at the speaker in the form of random noise. The amount of this noise depends upon the strength of the static signal. It may be strong enough to effectively drown out the audio signal.

The FM broadcast system does not depend upon amplitude modulation. As a result, the static or electrical noise in the atmosphere does not affect the received signal. Look at Fig. 9-8. This represents a typical FM

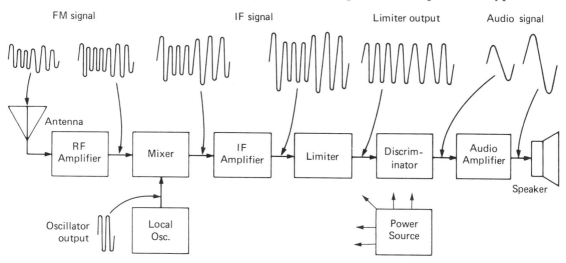

Figure 9-8. Signal processing in an FM radio.

receiver. A comparison of these blocks with those in Fig. 9-2 shows that one block has been added to the receiver. This block is called the *limiter*. Its function is to remove any possible amplitude change that might appear in the signal. The limiter block is found immediately in front of the detector. It presents a signal of uniform amplitude to the detector block. There are no other major differences between the signal carrying blocks of the FM radio and those of the AM radio. A comparison of the gain factors between these two radios shows that there is a great similarity. The amplitude of the waves between stages is about equal, stage for stage.

There are some significant differences inside the blocks of the radios. The type of detector system differs. This will be discussed in detail in a later chapter. Another difference is in the band of frequencies used for FM broadcast. This band starts at 88 megahertz and continues to 108 megahertz. A third difference is the IF frequency used for FM. This frequency is 10.7 megahertz. These three differences—the type of detector, the frequency range of the receiver, and the IF frequency—are important to the repair technician. When this information is understood, then the technician can understand how the system works and be better able to diagnose faults, in order to locate trouble areas in the set.

Frequency response.   Each type of radio receiver has limitations with regard to the audio frequencies received and reproduced. Chapter 7 discussed sidebands and bandwidth of the broadcast signal. AM radio stations are limited by the FCC to a bandwidth of 10 kilohertz. This means that the modulation information cannot exceed 5 kilohertz on either side of the carrier frequency. Since the width of the sideband is determined by the frequency of the modulation signal, the audio frequency response for an AM radio is limited to a range of 0 to 5 kilohertz. Audio notes higher in frequency than this 5-kilohertz cutoff-point are not broadcast.

The FM bandwidth is considerably greater than the AM bandwidth. It has been established at 75 kilohertz. This is well beyond the normal audio range of 20 hertz to 20 kilohertz. The bandwidth is large so that other information may be included on the station carrier. This information may include FM stereo signals as well as a *subsidiary carrier assignment* (SCA) for background music systems. When a carrier contains several different kinds of signals, the broadcast signal is classified as a *composite* signal. An analysis of the composite FM signal is shown in Fig. 9-9. The band of

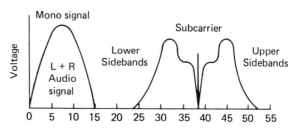

Figure 9-9. The composite FM-stereo signal.

frequencies from 0-15 kilohertz contains the regular FM non-stereo signal. This is the L + R (left + right) audio information, or sum of the signals in each channel. Audio signals up to a frequency of 15 kilohertz are transmitted. This is a much wider range of frequencies than in AM, hence a better quality audio signal is handled. Another signal (not shown) appears at 19 kilohertz. This signal contains no modulation. It is a reference signal for the stereo decoder section of the receiver. The group of frequencies from 23 to 53 kilohertz contain the stereo information. These are added to the carrier as a double-sideband suppressed-carrier signal. Both the upper sideband and the lower sideband contain the stereo information. This is labeled as L − R information. It is developed as a *difference* signal (left minus right). The suppressed-carrier frequency is 38 kilohertz. This frequency is exactly double that of the 19-kilohertz pilot-signal. The discussion of the stereo decoder action will show how this functions. The stereo information is added to the main carrier by a system called multiplexing. First, the information is amplitude modulated, and then it is added to the FM carrier signal. When the multiplexed information is received, it must be detected and reconstructed as audio information. This is done in the stereo decoder section of the receiver.

A simplified block diagram of an FM stereo receiver is shown in Fig. 9-10. One must assume that those blocks that exist between the antenna and the detector are still present. They have been omitted for purposes of clarity. Let's follow the composite FM-stereo signal through the balance of the receiver after it leaves the detector block.

We'll start by following the L + R signal. After this signal leaves the detector, it passes through a lowpass filter network. This network is designed to allow frequencies below 15 kilohertz to pass through it. It stops, or cuts off, higher frequencies. In this case, it would cut off the group of frequencies above 15 kilohertz, which includes all those frequencies containing the stereo information. The L + R signal then goes to a delay line. This delay line slows down the signal in order to allow the stereo signals to be processed in the decoder. Both the L + R and the L − R stereo information must reach the speakers at the same time if the original signal is to be reproduced. The delayed L + R signal then is sent to two blocks called *adders*. They will now meet the L − R signals.

Look again at the detector block. Two other signal paths are developed in this block. One of these paths is to a 19-kilohertz amplifier. This is a tuned circuit that is designed to amplify the 19-kilohertz pilot-signal and present it to an oscillator. The purpose of the pilot signal is to provide a reference signal for this oscillator. If the reconstruction of the suppressed carrier for the double-sideband L−R stereo-signal is to be accurate, then the oscillator has to be in step with the transmitted signal. The pilot signal is frequency multiplied to 38 kilohertz and used for the purpose of synchronization. It is also used to redevelop the 38 kilohertz carrier in order that the AM signal may be detected. While all of this is happening, the L−R signal also leaves the detector. It passes through another filter.

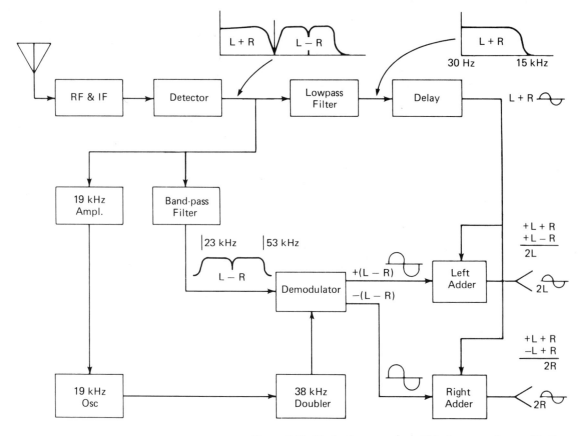

Figure 9-10. Block diagram of an FM stereo receiver, with the stereo-decoder section expanded.

This filter is called a *band pass filter.* It is designed to permit a limited group of frequencies to pass, and stop all others. The group of frequencies that pass this filter are those in the 23 to 53 kilohertz range that make up the L−R signals. The L−R signals and the 38 kilohertz oscillator signals are presented to the demodulator block. In this block the amplitude modulated information is removed from the carrier, or detected. One other thing happens to the signal. There are two outputs in this block. One signal is inverted at the outputs producing a + (L− R) signal and a − (L− R) signal. Simplifying these equations gives an L+ R and a − L+ R signal. Both of these are sent on to the adder blocks. In the left adder block the L−R from the demodulator is added to the L+R from the delay. This produces a 2-L signal, since the +R and the −R cancel each other. In the right adder block, the −L+R from the demodulator is added to the L+R from the delay, producing a 2-R signal. These signals, the 2-L and the 2-R, are then sent to separate channels of a stereo amplifier. They are amplified and converted into sound waves at the loudspeakers of the amplifier.

One other thing occurs in the stereo decoder block. This is the turning on of an indicator lamp, or device, so that the listener can see that the broadcast is in stereo. This device is usually a lamp. It is turned on by circuitry in the decoder. In some cases the 38-kilohertz oscillator voltage causes the lamp to light. Any of a variety of circuits can be used to indicate a stereo broadcast. The set designer and engineer choose one that is appropriate for the set.

9-3 Similarities in Systems

A great number of similarities exist among AM radio, FM radio, and FM-stereo radio receivers. Most of these were discussed in detail in Chap. 4. It would be more fruitful now to look at the differences in the signals as well as in how they are processed through the three kinds of radio receivers.

9-4 Differences in Systems

The chart shown in Fig. 9-11 will help you to see the obvious differences in the three receivers. Amplitude modulation is used for the AM receiver. Frequency modulation is used for both types of FM receivers. One assigned group of frequencies, or broadcast band, is used for the AM receiver. Another frequency group is assigned for use of FM broadcasting. Both FM and FM stereo use the same broadcast band.

Different IF frequencies are used. AM radio uses a frequency of 455 kilohertz. Some automobile radios use 262 kilohertz for their IF. This offers improved sensitivity and selectivity for the car radio. FM radios use a frequency of 10.7 megahertz for their IF frequency.

Bandwidth for AM is much less than for FM. The FM bandwidth for the monaural segment is limited to 15 kilohertz. This permits most of the audio frequency range of 20 to 20,000 hertz to be broadcast. The fidelity of FM broadcast is much greater than that of AM due to the increased bandwidth available for the signal. The section of the FM stereo band from 16 to 75 kilohertz is used for FM stereo, as well as SCA (subsidiary carrier assignment) used in background music systems. Although the

	AM	FM	FM Stereo
Modulation	Amplitude	Frequency	Frequency
Broadcast band	540–1,600 kHz	88–108 MHz	88–108 MHz
IF frequency	455 kHz or 262 kHz	10.7 MHz	10.7 MHz
Bandwidth	10 kHz	30 kHz	150 kHz

Figure 9-11. A comparison of kinds of radio receivers used for home entertainment.

bandwidth used for FM stereo is greater than that used for FM monaural, there is no increase in fidelity.

While the blocks have the same titles and perform the same basic functions, the frequencies of the signals used for AM and FM receivers are different. Actually, the differences occur in the carrier frequencies of the broadcast stations. The audio segment of the carrier uses the same frequency range of 20 to 20,000 hertz. Reception of the proper carrier-frequency is accomplished by tuning the receiver's RF amplifier and local oscillator. The frequency of the RF amplifier and local oscillator are dictated by the design of the receiver and the specific group of frequencies to be received.

SUMMARY   A method of adding intelligence to a broadcast station carrier-signal is utilized in order to provide a convenient means of communicating. The process of adding a signal in the audio range to a broadcast station carrier is called modulation. Many broadcast stations modulate the amplitude of the carrier. As a result, the carrier takes on the shape of the modulating wave. This method is one used by many broadcast stations. A group of frequencies is assigned for this use. It is called the AM broadcast band. The frequencies run from 540 kilohertz to 1.6 megahertz.

The type of receiver most commonly used today is called the superheterodyne radio receiver, or just *superhet.* It mixes the incoming signal from the broadcast transmitter with a signal produced in the local oscillator of the radio. The result of this action is a third signal called an intermediate frequency signal, or IF signal. Use of this system improves selectivity in the radio. Most of the amplification of the signal occurs in the IF amplifier block. After amplification, the signal is demodulated, leaving only the intelligence, or audio information. Further amplification occurs in the audio amplifier block. The output of this block is connected to the loudspeaker.

FM, or frequency modulation, was developed in an effort to overcome reception of noise and static. This system modulates the frequency of the carrier instead of the amplitude. As a result, static and noise, which are amplitude dependent, are not completely processed by the radio. The FM broadcast band covers the group of frequencies from 88 megahertz to 108 megahertz. The block diagram of an FM receiver is very similar to that of an AM receiver. Major differences are in an additional block, called a limiter, and in the frequencies received.

FM stereo receivers add an additional group of blocks. This group is generally called the stereo decoder section of the radio. In this section the stereo portion of the broadcast is amplified, detected, and added to the monaural portion. This creates both left-channel and right-channel information. The two channels of information are sent to separate audio amplifiers for further amplification and conversion into sound waves.

The similarities between these three receivers are great. The only basic differences are found in the group of frequencies received and in the additional blocks required for stereo reception and reproduction.

QUESTIONS

1. What two signals are present in a modulated signal broadcast from a radio transmitter?

2. What are the two basic modulation systems in use in home entertainment equipment?

3. What is the AM broadcast-band frequency spectrum?

4. What is the FM broadcast-band frequency spectrum?

5. Explain the action of the mixer block in the superheterodyne radio.

6. Explain the action of the detector block in a radio.

7. Which stage of the radio has the highest carrier gain? Which has the highest audio gain?

8. State the basic differences between an AM and an FM radio.

9. Where, in the FM station transmission, is the stereo portion of the broadcast?

10. List the blocks in the three systems (AM, FM, FM stereo) which are similar in action.

Signal Paths: Television Receivers

10

The television receiver as it is known today is a very complex device. As is true with most complex devices, the TV receiver may easily be analyzed into smaller blocks. Each block has its function, as does the whole set. The blocks are interrelated as far as their individual functions are concerned. Look at Fig. 10-1. This drawing represents a TV receiver that has been reduced from a large block containing input and output devices into smaller intermediate blocks. This figure illustrates the various signal paths in the receiver. A TV receiver must accomplish three basic things: (1) reproduce a picture with the proper brightness and color; (2) reproduce the proper sound to accompany the picture; and (3) synchronize the picture, sound, and color with that sent by the TV station transmitter. Much information has to be transmitted in order to accomplish these three functions. A review of the signal known as the composite video signal is in order at this time.

10-1
The Composite
Video Signal

The composite video signal is shown in Fig. 10-2. This signal contains all of the required picture and synchronizing information necessary to display a picture on the CRT. The signal is composed of picture information, blanking information, and sync information. The voltage or current levels relate to the brightness or darkness in the picture. Industry standards indicate that the black level is 75% of the maximum signal amplitude. Therefore, it is possible to transmit information when the screen

134

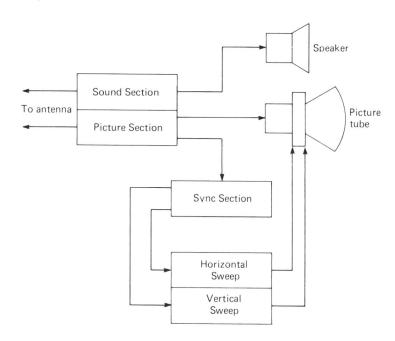

Figure 10-1. Modified block diagram of a TV receiver, showing signal paths in the receiver.

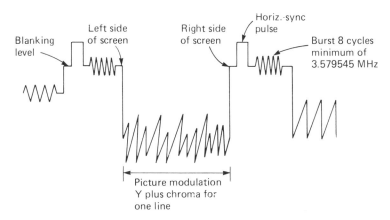

Figure 10-2. A line of composite video information. Note the placement of picture, blanking, color-burst, and sync signals.

of the CRT is black. Most of the sync information is transmitted at this *blanked,* or blacked out, signal level. Note that the picture information varies greatly as the intensity of the picture varies.

Refer to Fig. 10-3. This reference is for a multi-line signal. One horizontal blanking-and-sync pulse is transmitted for each line of the picture. At the top of the field, several lines of information are transmitted when

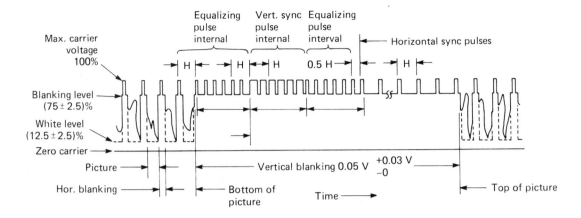

Max. carrier
voltage
100%

Equalizing
pulse
internal

Vert. sync
pulse
internal

Equalizing
pulse
interval

Horizontal sync pulses

H

H

H

0.5 H

H

Blanking level
(75 ± 2.5)%

White level
(12.5 ± 2.5)%

Zero carrier

Picture

Vertical blanking 0.05 V

+0.03 V
−0

Hor. blanking

Bottom of
picture

Time

Top of picture

Figure 10-3. The composite video signal and the sync information sent during the vertical blanking interval at the bottom of the picture.

the picture is blanked. This information is related to the vertical sync and is transmitted when the picture is blacked out. This is done for 21 lines of each picture field. The composite video signal has to be received and amplified through the TV set so that the proper picture may be displayed on the CRT.

10-2
Signal Paths

The block diagram in Fig. 10-4 is that of a typical black-and-white TV set. The waveform at the antenna represents a composite video signal, as it has modulated the RF carrier. The amplitude of this signal is on the order of 50 microvolts (0.00005 volts). This is the level of signal required to produce a signal at the CRT of 50 volts. A signal of 50 volts at the CRT is a minimum requirement for a quality picture. The composite signal is received in the RF amplifier section of the tuner and amplified. It is then sent to the mixer block. At the same time this is happening, the local oscillator is generating its own carrier signal. In the case of the TV set, the local oscillator normally operates at a frequency above that of the received signal. The RF signal and the oscillator signal are heterodyned in the mixer block. The result of this action is a modulated composite-video signal with a carrier frequency of 45.75 megahertz. The 45.75-megahertz frequency is used as the IF frequency in TV receivers used in the United States. The FM sound signal rides along with the composite video signal. Even though it has also changed its frequency, there still is a difference of 4.5 megahertz between the IF carrier and the sound carrier. This difference of 4.5 megahertz is the same as was used on the original broadcast transmission. The signal has been amplified to about 120 times its original value at this point in the set. This increases the amplitude of the signal to a level of about 0.006 volts.

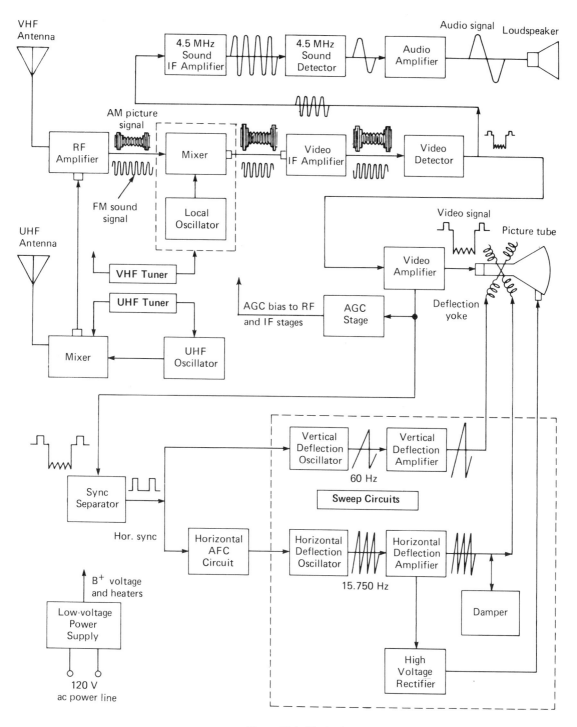

Figure 10-4. Block diagram of the TV receiver, with all signal-information waveforms displayed.

137

The next block through which the signal passes is the IF amplifier. This is the block in which most of the amplification occurs. The amount of amplification is quite large. Typical gain figures for an IF amplifier will produce a signal that may be as much as 700 to 800 times larger than the input signal to this block. An amplification factor of 700 is used in the TV set under discussion. The result of this amount of amplification is a signal level of about 4.2 volts. The original signal level at the antenna terminals of the TV set has been amplified from 50 microvolts to 4.2 volts. This represents a gain figure of 84,000 times the input signal at this point in the set.

The next process for the signal is detecting. In the detector block, the signal is literally cut in half horizontally, permitting only half of the 4.2 volt signal to pass through the detector block. The result of this action produces a loss of signal and an amplitude of about 2.1 volts. The detector also is used to eliminate the IF carrier at this point, leaving only the amplitude modulated waveform of the video information.

At this point in some sets the sound and sync information is removed. Other sets perform this operation at the output of the video amplifier block. The sync and sound signals are sent to other blocks for further processing. Let's finish with the picture signal path before these other signals are discussed.

The signal, with an amplitude of about 2.1 volts, is now sent to the video amplifier block in the set. Further amplification occurs in this block. The end result is a signal level of about 50 volts. The amplification factor has to be about 25 in order to produce this level of signal. Total signal gain in this TV receiver is over 2,100,000 times the input signal. This signal is sent to the CRT where it is used to control the electron intensity in order to reproduce the original image viewed by the TV camera. The signal is normally sent to the cathode of the CRT. Variations in signal level are directly related to degrees of lightness or darkness on the screen.

The sound signal path.   While all of this is occurring, the sound signal is being processed. The sound system used in today's TV set is called an *intercarrier sound system*. Using this kind of system permits both the sound and the picture information to be tuned at the same time with only one selector control. The frequency of the sound carrier is always 4.5 megahertz away from the picture carrier frequency. The sound is heterodyned in the tuner mixer and is amplified through the IF amplifier. At the detector, another mixer action occurs. Here, the picture IF frequency is heterodyned with the sound IF frequency in order to produce another IF frequency, of 4.5 megahertz. A tuned circuit called a *sound trap* is employed in the set at this point. This trap passes the 4.5-megahertz FM sound-signal to the sound IF and rejects all other frequencies. As a result, a sound IF signal at a frequency of 4.5 megahertz is sent to the sound IF amplifier block. In this block it is amplified and sent to a sound detector block. In a color set the video and sound signals are both sent to a sound detector. Here the video is trapped out and the sound passes on to the sound IF block. This is done in order to minimize interaction between the 4.5-megahertz sound-IF carrier frequency and the 3.58-megahertz color-subcarrier frequency in the detected video signal. Interaction between these two frequencies in the video section of the TV set produces a herringbone interference pattern in the video.

From this point on, the sound signal is processed in a manner similar to that in the FM radio receiver. Detector action separates the information from the carrier, leaving only the audio signal. The audio signal is amplified through blocks working as voltage amplifiers and power amplifiers, and is then sent to the loudspeaker for conversion into sound waves. Gain figures for sound signals compare to those used in other audio devices.

The sync signal path. The sync signals are also amplified through the set. They, too, are removed at the output of the video detector block. The exact point from which both the sync and sound signals are sent to their respective blocks will vary with different sets. The technician reads the schematic diagram for the specific set in order to find these points. The sync signal is sent to a sync separator block. The signal is often amplified in this block. There may be either one or two outputs from the sync separator, depending upon the set's design. The separator is designed to react only to the sync pulses sent from the transmitter. The other information on the composite video signal is ignored by this block. The output signal is sent to both the vertical oscillator block and the horizontal AFC, or oscillator control, block.

The sync signal is used as a timing signal. It literally tells both the vertical oscillator and horizontal oscillator when to start, in order to keep both oscillators in phase with the broadcast information. When the sync signal is not present, the picture will either roll vertically or pull horizontally, or both. Figure 10-5 depicts these conditions.

Figure 10-5. A TV picture with loss of both vertical and horizontal sync.

Scanning. In Chap. 2 we discussed the carthode-ray tube and the movement of the electron beam. The electron beam in a TV set's CRT sweeps across the face of the tube horizontally, one line at a time. After each sweep, the beam is repositioned to sweep across another line. U.S. standards for TV call for a picture composed of 525 horizontal lines. In order to improve viewing quality, the horizontal sweep works on an every other line principle. This means that all of the odd-numbered lines are swept, one at a time, and then the electron beam is positioned to sweep all of the even-numbered lines. The total sweep pattern makes up a complete picture of 525 lines. This process of sweeping is called *scanning.* The process of interweaving the sweep is called *interlace scanning.* The movement of the electron-beam position is controlled by two sets of electromagnets. These are in the deflection yoke, which is placed around the neck of the CRT. One set is connected to the horizontal sweep system. The other set is connected to the vertical sweep system. Both sets have to be working at the same time in order to force the beam to move both horizontally and vertically. Only two vertical scans are made while the beam is scanning the 525 horizontal lines. Because of this, the vertical sweep rate is much slower than the horizontal sweep rate in the TV set. Each set of scanned lines makes up a field of 262-1/2 lines. This is illustrated in Fig. 10-6. Look at the field for the odd-numbered lines. The *trace* starts at the top of the picture. It moves across the face of the CRT and then is repositioned back at the starting side, but at a lower level. This procedure is repeated for all 262-1/2 odd-numbered lines in the field. Note also that when the trace reaches the bottom of the screen it returns to the top, the center this time, to start again. A complete scanning procedure for both the odd and the even fields is called a *frame.* There are 30 frames a second in a complete picture. The period of time required to return the beam to the proper side of the CRT screen is called the *retrace time.* This period is extremely short when compared to the scanning time. During this time the beam is turned off, or blanked out.

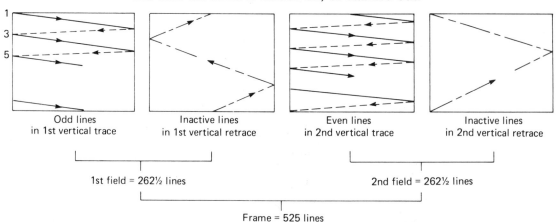

1
3
5

| Odd lines in 1st vertical trace | Inactive lines in 1st vertical retrace | Even lines in 2nd vertical trace | Inactive lines in 2nd vertical retrace |

1st field = 262½ lines   2nd field = 262½ lines

Frame = 525 lines

Figure 10-6. Scanning lines during trace of each field.

The activity called scanning is constantly going on in an operating TV set. The sweep section in the TV set contains circuitry for producing both the vertical and the horizontal sweep patterns. The required waveforms are generated by oscillators. The vertical oscillator block generates a signal at a frequency of 60 hertz. The form of this signal is sawtooth. The sawtooth waveform is used to develop a linear, or constantly rising, change in deflection-yoke current, so that the spacing between each line is constant. A nonlinear sweep-wave would produce a distorted picture—either a compressed top or bottom of the picture, or possibly an expanded picture on the top or bottom of the screen. None of these distortions is desirable. Look at Fig. 10-7. It shows the blocks related to the sweep sections of a TV set. In the vertical sweep block, a sawtooth wave is produced. The signal is amplified in the vertical amplifier-output block and sent to the vertical set of coils in the deflection yoke. The result of this action is a controlled vertical movement of the electron beam in the CRT.

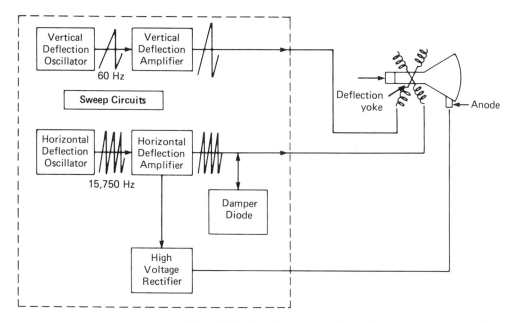

Figure 10-7. Waveforms associated with the sweep sections of the receiver.

While this is happening, the horizontal oscillator is also generating a sawtooth wave. Its frequency is about 15,750 hertz. This frequency is produced when 525 lines are required at a rate of 30 times a second. ($525 \times 30 = 15,750$). In a color set, the frequency is slightly less than 15,750 hertz. There will be more on that later in the chapter. This signal, like that of the vertical sweep section, is amplified in the horizontal amplifier-output block. It is then sent to the horizontal coils of the deflection yoke. The result of this action is a controlled horizontal sweep of the electron beam in the CRT.

Picture elements. Another item important to picture composition is the amount of brightness produced when the electron beam strikes the phosphor coating on the inside of the CRT screen. There are about 144,000 useable phosphor spots on each CRT. Each one of these spots makes up an element of the total picture. A strong electron beam makes the spot glow brightly. A weak beam will produce little, or no, glowing effect. The strength of the beam is controlled by the amount of signal information received and processed by the TV set. This is directly related to the level of signal contained in the video information section of the composite video signal.

10-3
Color Signal Paths

One of the major problems relating to the color TV faced by the electronics industry was that of compatibility. The FCC insisted upon a color system in which any color picture broadcast could be received and viewed on a black-and-white TV as a black-and-white picture. A system was developed by which this could be accomplished. This is the system which is in current use. The color information is added to the main signal carrier as a multiplexed signal. Use of this kind of system allows for both color and non-color broadcasting. It also permits all TV sets to receive and process both color and non-color signals. Naturally, a non-color TV set does not process the color signal information, but reproduces, instead, a black-and-white picture. The diagram in Fig. 10-8 shows how this is accomplished.

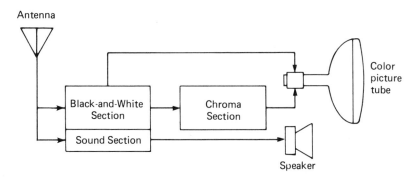

Figure 10-8. Modified block diagram of a color TV receiver.

The color-set block diagram is developed from the basic black-and-white set. An additional section, called a *chroma* section, has been added to the basic blocks. The signal from the video amplifier is fed to this large block as well as to the CRT. In other words, there are now two paths for the signal to take from the video amplifier to the CRT. Before these paths, and the additional blocks required to process the color information, are discussed, a look at the color signal's composition and the color CRT is in order.

The color signal. Color information as used in a color TV set is composed of three colors: red, green, and blue. These colors are used because they are the colors required to produce all colors. Other colors, including white, are produced by combining the proper amounts of these three primary colors. Three video color-signals are produced by the color TV camera. These three colors are mixed to produce a signal that corresponds to the amount of lightness or darkness in the picture. This signal is called the Y signal. It is very similar to the modulating signal used to produce a black-and-white picture on a non-color set. At the same time this is occurring, the color information is being amplitude modulated onto a 3.58-megahertz carrier. Actually, two signals are being developed in this manner. One signal represents a green-purple color signal and is called the Q signal. The other signal represents an orange or cyan-blue color signal and is called the I signal. These two signals are added to the 3.58-megahertz color carrier 90 degrees out of phase with each other. In this way, they are transmitted without interference with each other. The Q and I signals combine at the transmitter to form the C, or chrominance, signal. When these signals form the C signal, the 3.58-megahertz color carrier is suppressed. The result is a double-sideband suppressed-carrier signal carrying all of the color information required to make up a color picture. Both the C, or color, signal and the Y, or luminance, signal are used to redevelop the total color picture in the TV set. These signals are amplitude modulated onto the broadcast station carrier in order to form the composite *colorplexed* video-signal.

You may recall that the FM stereo-subcarrier signal had to be re-created in the receiver before demodulation could occur. It was necessary to provide a pilot signal on the carrier in order to synchronize the station and the receiver. This is also true with the color signal. It, too, is a suppressed carrier signal. The receiver must be able to re-create the 3.58-megahertz signal in order for demodulation to occur. A sync signal for color must also be transmitted with the other information. A place in the broadcast signal had to be found which would carry this information without interfering with any other information. The *color-burst* signal is placed on the *back porch* of the horizontal blanking pulse. A minimum of eight cycles of this signal are sent with each horizontal pulse when a color signal is being transmitted. The purpose of this color-burst signal is to synchronize the color carrier in the receiver and to turn on the color processing blocks of the set. The composite video signal in Fig. 10-2 shows the position of the color-burst signal.

The color CRT. Several methods of displaying color on the face of the CRT have been developed. One method has become more or less standard in the United States. This has evolved into a color CRT called the *shadow-mask* CRT. It was mentioned earlier that each of the color tube's phosphor dots is actually made up of three dots, one for each of the primary colors of red, green, and blue. There are three electron guns in the neck of the CRT, one for each color. The electron beams from each gun must be focused on one set of dots in order to reproduce the correct color on the screen. This is shown in Fig. 10-9. The three electron beams are made to pass through the same hole in the shadow mask. They then spread apart slightly in order for each to land on the correct color dot. There are several types of color CRTs in use today. Some CRTs use phosphor bars instead of dots. Others use a variety of gun designs. Regardless of its type of construction, each has to get the proper electron beam and the proper intensity onto the right spot on the face of the CRT in order to reproduce the color picture.

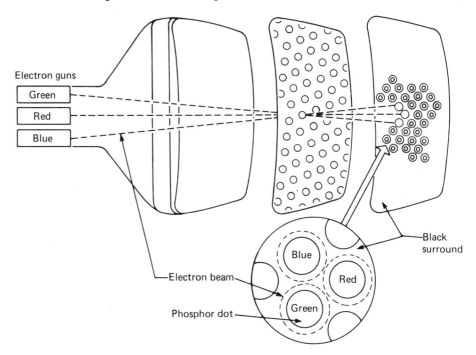

Figure 10-9. The color CRT, shadow mask, and screen. Note the triad of color dots.

The color blocks. There are several blocks which make up the chroma section of a color TV set. These blocks are illustrated in Fig. 10-10. The C signal from the video amplifier block is sent to a chroma-bandpass amplifier block. This block is designed to pass only those frequencies related to the color signal. These frequencies are clustered around the 3.58-megahertz suppressed-carrier frequency. The color information is amplified and sent to the color demodulator blocks. There are several systems in use currently for color demodulators. Since our intention is to show the basic signal paths, rather than to present specific circuit descriptions, only a basic block system is considered here. Regardless of the system used in the set, there is a demodulator for each of the primary

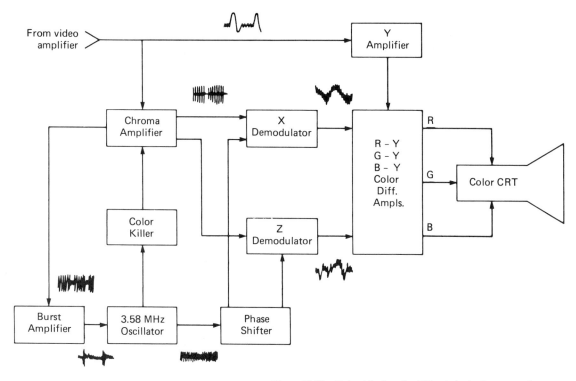

Figure 10-10. Color blocks of a TV set, including waveforms.

colors, in most sets. Since these demodulators are working only with the color information and not with the luminance, or Y, signal they are referred to as X and Z demodulators.

The color demodulators are unable to separate the color information from the suppressed-carrier signal until a carrier is available in the set. The color oscillator used as a reference in the set has a frequency of 3.579545 megahertz. This is rounded off to 3.58 megahertz for convenience. The color-burst signal riding on the back porch of the horizontal sync and blanking pulse is used as a reference signal in the color processing blocks. This signal really performs two functions in the set. One is to control the phase of the color oscillator. The color-oscillator signal produced in the set has to be in phase with the transmitted signal in order to reproduce accurate colors in the receiver. The color burst acts as a reference for this circuit. Note the waveforms in Fig. 10-10 as they relate to the oscillator block. Both the color-burst and color-oscillator signals must start at the same exact moment.

The other function of the color-burst signal is to control a color-killer block. The burst signal, when present, turns off the color killer. The output of this block feeds into the chroma-bandpass amplifier block. When no color burst is being transmitted, the color-killer block turns off the bandpass amplifier. This in effect turns off all of the color circuits in the set. The result is a black-and-white picture on the CRT.

The output of the color-oscillator block is fed into the X and Z demodulator blocks. The color oscillator re-creates the 3.58-megahertz carrier which was suppressed at the transmitter. This carrier is required in order to demodulate the color information. Only two demodulators are required in this system even though there are three basic colors used in the TV system. The output of the X demodulator is the $R-Y$ signal. The output of the Z demodulator is the $B-Y$ signal. So far, two of the three colors have been demodulated. These are sent to the color video amplifiers. The third color, $G - Y$, is mathematically related to the $R - Y$ and $B - Y$ signals. It is re-created by combining these two signals in an amplifier stage. The three color-signals are then sent to the color CRT where they are combined with the Y, or video, information in order to re-create the full color picture.

A review of the color processing blocks in the TV set shows that there are three signals required in order to display a color picture. The first signal is the Y, or luminance, information. This is sent from the video amplifier to the three color-video amplifiers. It is necessary to slow down this signal slightly in its journey because of the additional time required to process the color information. By so doing both the color and the luminance signal arrive at the color video-amplifiers at the same moment. A device called a *delay line* is used to slow the speed of the signal in the Y path.

The second signal is the C, or color, information. This is sent from the video section of the set to a chroma-bandpass amplifier block, where it is amplified and sent to the color demodulators. From these two blocks a third color is produced. The three colors are then sent to color video (or *color difference*) amplifiers. These blocks are identified as $R - Y$, $B - Y$, and $G - Y$ (red minus luminance, etc.). After leaving these blocks the signals are sent to the color CRT.

The third signal is the color burst. The purpose of this signal is to act as a reference for the color oscillator in the set. It also turns off the color-killer block when a color signal is being transmitted.

The color oscillator-signal is used to create the 3.58-megahertz color carrier that had been suppressed at the transmitter. This is required in order to demodulate the color information. The output of the color oscillator is sent to the demodulators. The demodulator signals are sent to three color-video amplifiers. In each of these amplifiers, the information related to the specific color (red, green, or blue) is combined with the video information and sent to the color CRT, where it is used to display a picture.

There are several types of color processing sections in use today. The exact circuitry and block setup will vary among set manufacturers. The method of processing the color information does not vary too greatly in any case. The material presented in this section will assist you in understanding how the color information is processed in most sets.

The key word in describing the similarities between black-and-white and color TV systems is compatibility. All TV sets must be able to receive both color and non-color broadcasts. The non-color set reproduces a black-and-white picture regardless of the kind of signal transmitted from the station. The color set has additional circuitry so that it is able to reproduce either non-color or color broadcasts in their original form.

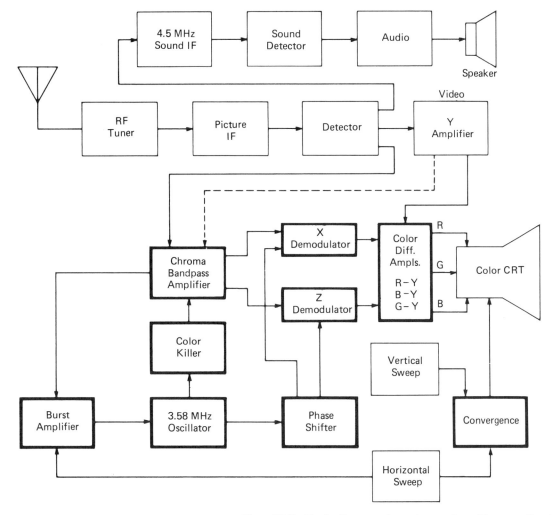

Figure 10-11. Block diagram of a color receiver. Heavy-outlined blocks are used only for color reproduction.

Figure 10-11 shows the block diagram of a color receiver. Many of the blocks used in this set are common to both color and non-color receivers. Both systems have similar signal-processing blocks from the RF amplifier in the tuner to the video amplifier. Both systems use the same kind of horizontal and vertical sweep blocks. The signals and waveforms found in the blocks mentioned are similar in both the color and non-color receiver.

**10-5
Differences
between TV
Systems**

The major differences between color and non-color receivers are found in the color processing blocks and in the CRT. An entire additional section, specifically designed to process and decode color information, is added to the color receiver. The CRT now has to be able to develop three separate colors, in the proper proportion. These colors have to be mixed and displayed on the CRT in order to re-create the color picture. This requires a CRT that is capable of generating three electron beams, one for each of the three primary colors. Also, higher operating voltages are required for this CRT. The necessity for higher operating voltages introduces still another problem. The higher voltage has to be maintained at a constant value. This means that a voltage regulating circuit is required, in addition to the other blocks not found in a black-and-white TV. Variation in voltage at the CRT produces variations in color. This problem is overcome when the *regulator circuit* is added to the set. These blocks are all displayed in Fig. 10-11. The darker blocks represent those that are used only in a color receiver. Another section of the color set, which has not been discussed previously, is the *convergence section.* This unit contains the necessary circuitry to enable each electron beam to land on the proper phosphor dot anywhere on the face of the CRT. When the convergence section is properly adjusted, the color picture is clear and sharp; otherwise, the picture may have shadows of color outlining the images on the screen. All other blocks are common to both color and non-color receivers.

SUMMARY

A functional TV set must do three things: reproduce a picture of the proper brightness and color; produce the proper sound; and synchronize picture, sound, and color with that broadcast by the station. The composite video signal contains all of the necessary picture information. This signal is developed in such a way that one line of information is transmitted at a time. The picture is scanned, one line at a time, by an electron beam in the CRT. Because the human eye tends to retain images, the individual lines appear as a complete picture.

The TV set consists of several major blocks. These blocks include video signal processing, audio signal processing, synchronizing, horizontal and vertical sweep, and color processing. Each block is designed to accept the set of electronic signals it is designed to process and to reject all other signals. Most of these blocks are amplifiers, so that the signals are enlarged as they are processed.

The tuner section of all TV sets is similar to the tuner section of the radio. It contains an RF amplifier, a mixer, and a local oscillator. The operating principle is the same as for a radio, but the operating frequencies are very different. The output of the mixer is a signal whose frequency is 45.75 megahertz. This is the picture intermediate frequency for all TV sets. Other standard frequencies related to this IF are 3.579545 megahertz, used as the color oscillator frequency, and 4.5 megahertz, used as the sound intermediate frequency. These last two signals are developed in the transmitter.

After passing through the mixer, the composite IF signal is amplified to about 700 or 800 times its initial level. It is then detected and sent to the audio, video, color, and sync sections of the set for further processing.

Another section of the set is the sweep section. There are both horizontal and vertical components in the sweep. Each has its own operating frequency. Horizontal sweep is at a rate of 15,750 hertz. The vertical sweep rate is 60 hertz. The horizontal sweep also helps develop the high voltages used in the set. The sweep section produces the 525 lines of the picture as well as the total picture *frame*. Picture information is developed in the video section and added to this frame in order to reproduce the signal as seen by the TV camera at the transmitter.

Color information is also developed at the transmitter. It is processed in the chroma section of the receiver. A signal called the color burst is transmitted, along with other information, during a color broadcast. The burst is used to turn on the chroma sections and to synchronize the color oscillator. During a black-and-white broadcast the color burst is not transmitted. The color processing blocks are only on when the burst signal is received. They are turned off by a color-killer section when a black-and-white signal is received.

The basic difference between a color receiver and a non-color receiver is the lack of color processing blocks in the latter set. Other than this, the signal processing is the same. Compatibility was one of the major requirements established by the Federal Communications Commission when the possibilities of color-signal transmission were being developed. As a result, we now are able to receive the same basic picture information on both types of TV set.

QUESTIONS

1. Name the five major blocks of a black-and-white TV set.

2. What are the two purposes of the color-burst signal?

3. Draw a block diagram of the video-signal processing blocks.

4. State the following frequencies:
   (a) video IF
   (b) color
   (c) audio
   (d) vertical sweep
   (e) horizontal sweep

5. What is the purpose of the sync signal?

6. Which blocks of the TV are controlled by the sync signals?

7. What is the basic wave-shape of the scanning signals?

8. What are the three basic colors used in a color TV set?

9. What is the purpose of the color oscillator block?

10. Why is it necessary to use a color demodulator?

11. Name the color processing blocks of a TV set.

12. Is it possible to display a black-and-white picture on a color TV set? Explain your answer.

Section Four

Components
and
Circuits

Electronic
Components

Many electronic terms were introduced in earlier chapters. The purpose in doing this was to familiarize the reader with these common terms and how they are used in the world of electronics. There are very few basic theoretical units to learn in electronics. In fact, there are only three types of basic *components:* resistors, inductors, and capacitors. These components are produced in a wide variety of shapes and values, but basically each group uses the same theoretical base. Along with the three components, there are three *concepts* which are considered basic. These are resistance, voltage, and current. One might consider power as another concept. However, since electrical power is a product of voltage and current, it should be considered to be a result of the other concepts instead of one of the basics.

Three kinds of electronic *circuits*, which are also basic, are the amplifier, oscillator, and rectifier. These also come in a variety of configurations. Here, as in the other fundamental things, the basic principles apply to all amplifiers, etc. The technician learns to identify each kind of circuit. He also learns how each variety of circuit is supposed to work. Three other items are part of the basic building blocks of electronics. These are the types of *circuit configurations* commonly found in electronic devices. The circuits are identified as series, parallel, or combination series-parallel.

Familiarity with each of the 12 things mentioned is considered fundamental to an understanding of electronics. They are "must" knowledge for those who are in the field of repair. Many technicians know how each of the components found in electronic devices functions. Some seem to have problems relating the functions to a complete circuit. Successful technicians know the device, as well as how it operates in an electronic circuit. In the next four chapters we will discuss each of these twelve concepts. The first part of this presentation is directed toward the theory behind the basic components as well as some of the interactions that take place when these components are combined in electronic circuits.

This book studies the electron in action. The action referred to here concerns the behavior of the electron in various electronic circuits. Each atom of every kind of material contains electrons. The number of electrons depends upon the electrochemical composition of the material. The ability of electrons to move freely in the material is also dependent upon the kind of material involved. Some materials, such as silicon, mica, glass, and rubber, do not allow free movement of electrons. These materials are called *nonconductors*. They are often referred to as insulators. Other materials, including silver, copper, and aluminum, have many electrons, and some of these electrons are reasonably free to move about in the material. Such materials are classified as *conductors*. They have the ability to permit transfer of electrons from one place to another place with little opposition to this movement. Some materials are not true insulators nor are they true conductors. Their ability to allow electron movement lies between the two extremes. These materials are called *semiconductors*. There is more about semiconductors in a later chapter.

The comparative ease or difficulty of movement of electrons within the material is called the *conductance* of the material. Conductance is rated in *mho*'s. (Mho is ohm, spelled backwards.) This term is not one of the more commonly used terms in electronics. It should, however, be recognized for what it is in the technical literature available. It refers to the ability of a substance to permit flow of electrons and the ease with which this is done. A more commonly used term in any discussion of electron movement is *resistance*. This term is the opposite of conductance. It refers to the opposition to electron movement in a circuit. A look at the properties of resistors is in order at this time.

11-1
Resistors

Most resistors used in electronics are made from either a carbon compound or a resistance wire. Resistors are classified in two ways. One way refers to the actual amount of resistance in the device. This is the ohm rating of the resistor. The second kind of rating is according to the amount of power the resistor is capable of handling. This is the watt rating of the resistor. Both ratings are important when selecting a resistor for use in a specific circuit.

Power ratings. First, let's look at the power ratings of resistors. This rating refers to the amount of heat the resistor is able to dissipate into the air and still perform its work. A product of resistance is heat. Opposition to electron movement in a resistor produces heat. This energy is not used in the circuit and must be vented into the air. Failure to accomplish this often results in an overheated resistor. Overheating may cause a fire or a burned-out resistor. In the process of overheating the resistor's ohmic value often increases beyond that intended by the manufacturer. This causes changes in circuit operating conditions which often cause the electronic device to stop working correctly.

Resistors are divided into two categories: *power resistors* and *composition resistors*. The dividing point is based upon the power rating. This point lies between two and three watts. Resistors rated at less than three watts are normally made from a carbon compound, and are called composition resistors. These resistors are molded, and leads are attached to the ends of the molded carbon form. Then the whole thing is encased in a plastic housing. The housings come in standardized sizes. Each standard size relates to a wattage rating. Higher wattage resistors require more surface in order to dissipate the larger amounts of heat. Because of this, the size of the resistor case is a good indicator of the wattage rating of the unit.

Resistors rated higher than two watts are called power resistors. They are manufactured by winding a piece of *resistance wire* on a ceramic form. These resistors are called wire-wound resistors because of the method of construction. Here, as with the carbon resistors, the physical size of the resistor is an indicator of the power rating. Power resistors may be obtained with ratings from 3 watts all the way up to 200 watts, or more.

Wattage ratings for any resistor are determined by applying *Watt's law*. In one form, this law states that power is equal to the product of the voltage times the current. If one knows these factors then it is easy to find the power rating required for a specific resistor. For example, a certain resistor has ½ ampere of current flowing through it. There is a voltage of 16 volts developed across the resistor. Using Watt's law of P (power) = E (voltage) × I (current) and substituting known values for the letters, then $P = 16 \times 0.5$, or 8 watts of power. In this case, a power resistor rated at 8 watts or more would have to be used in the circuit.

Resistance ratings. All resistors, regardless of their wattage ratings, are rated in *ohms*. This tells the specific amount of resistance in the device. In general, the higher the ohmic value, the greater the resistance in the resistor. In theory, a resistor could be built that had an infinite amount of resistance. In practice this is not true. An infinite resistance would act like an open circuit in the set and no current could move through it. The resistor is marked with its ohmic value when it is made. There are two methods of doing this. On power resistors the value is printed, with ink or paint. All one has to do is to read the numbers printed on the body of the resistor in order to determine its value.

Carbon resistors use another system of marking ohmic value. This system uses an industry-wide standard color-code. The resistor has at least three bands of color on it. The color bands are placed on the body of the resistor, close to one end. Each color refers to a number. Figure 11-1 shows how this is done. Start at the color band closest to the end of the resistor. This band represents a number. The exact number is determined when the chart is used. The second band also relates to a number. This is always true. Regardless of what happens after this point, each of the first two bands always refers to a specific number identified by the color. The third band is used in a different manner. It still relates to the value of the resistor, but, for this band, and only this band, we are looking at a multiplication factor. Another way of stating this is that this band tells the number of zeros in the value of resistance. For example, the color orange, in the code, refers to the number *three*. If this color is used in either or both of the first two bands, then the number *three* is identified. If the third band is orange, then the first two numbers are multiplied by 1,000, that is, three zeros are placed after the first two numbers. An example of this is a resistor with the color code bands of red, purple, and orange. On the chart, red refers to the number 2, purple to the number 7 and orange, in the third band, to either *000* or × *1,000*. We see that the resistor's color-coded value is 27,000 ohms.

The EIA Color Code			
Color	Meaning of color		
	In Band I or II	In Band III	In Band IV
Black	0	Add no zeros (× 1)	—
Brown	1	Add 1 zero (× 10)	—
Red	2	Add 2 zeros (× 10^2)	—
Orange	3	Add 3 zeros (× 10^3)	—
Yellow	4	Add 4 zeros (× 10^4)	—
Green	5	Add 5 zeros (× 10^5)	—
Blue	6	Add 6 zeros (× 10^6)	—
Violet	7	Add 7 zeros (× 10^7)	—
Gray	8	Add 8 zeros (× 10^8)	—
White	9	Add 9 zeros (× 10^9)	—
Gold	—	Divide by 10 (× 10^{-1})	±5%
Silver	—	Divide by 100 (× 10^{-2})	±10%

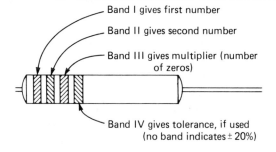

Band I gives first number

Band II gives second number

Band III gives multiplier (number of zeros)

Band IV gives tolerance, if used (no band indicates ± 20%)

Figure 11-1. The EIA resistor color-code system.

Some resistors have a fourth color band. This refers to the tolerance value. There are only three tolerances commonly used. These are ±20%, ±10% and ±5%. This tolerance establishes a set of upper and lower limits for the ohmic value of the resistor. For example, a resistor with a fourth color band of silver would have a tolerance of ±10% of the marked value. A 100-ohm resistor with a ±10% tolerance could range in actual value between 90 and 110 ohms and still be acceptable. A missing tolerance-band indicates a tolerance of ±20%.

A fifth color band is sometimes used on resistors. This band identifies a *reliability rating*. It is more commonly found on resistors used in government or military equipment. Actually, this is a failure rating; it tells us how many resistors of this type will fail after 100 hours of use. This information is not required for home-entertainment equipment at this time.

Standard values.   Over the years, there has been an effort on the part of manufacturers to standardize electronic components. One area where this has met with much success is in the agreement on a set of standard values for low-wattage resistors. Figure 11-2 shows how these standard values are arranged. Although the chart shows only a set of values ranging from 10 to 100 ohms, these values can be extended from fractions of an ohm all the way up to several million ohms. The multiplier, or

20% Tolerance	10% Tolerance	5% Tolerance
10	10	10
		11
	12	12
		13
15	15	15
		16
	18	18
		20
22	22	22
		24
	27	27
		30
33	33	33
		36
	39	39
		43
47	47	47
		51
	56	56
		62
68	68	68
		75
	82	82
		91
100	100	100

Figure 11-2. Standard values for resistors, in ohms.

157

number of zeros, changes while the first two numbers remain as shown in the chart. Low-wattage resistors used in electronic devices will use one of these values for a standard ohmic value. More numerical values are available as the tolerance becomes more exact and is reduced from 20% to 5%. Critical circuits will use 5% tolerance resistor values more often than noncritical circuits.

Another area of standardization is in the physical size of carbon-composition resistors. The size relates to the wattage rating. Most carbon resistors are rated in either ¼-, ½-, 1-, or 2-watt sizes. Figure 11-3 is a photograph of several carbon resistors. The wattage rating is directly related to the size of the resistor. As the power rating increases, the size of the body also increases. Keep in mind that the power rating has absolutely no effect on the ohmic value. It is possible to obtain a 100-ohm resistor rated at each of the standard power values.

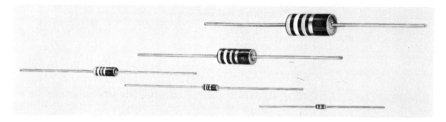

Figure 11-3. Standard sizes for carbon resistors. (Courtesy Ohmite Mfg. Co.)

A kind of shorthand is used for electronics measurements, as well as other scientific measurements. This shorthand uses a set of abbreviations which replace the number of zeros used in values. This applies to all components—resistors, capacitors, and inductors. It becomes more convenient to use the terms instead of writing all of the zeros which would otherwise be necessary. These terms are shown in Fig. 11-4. Note that

Value	Prefix	Symbol	Example
$1,000,000,000,000, = 10^{12}$	tera	T	$THz = 10^{12}$ Hz
$1,000,000,000 = 10^{9}$	giga	G	$GHz = 10^{9}$ Hz
$1,000,000 = 10^{6}$	mega	M	$MHz = 10^{6}$ Hz
$1,000 = 10^{3}$	kilo	k	$kV = 10^{3}$ V
$100 = 10^{2}$	hecto	h	$hm = 10^{2}$ m
$10 = 10$	deka	da	$dam = 10$ m
$0.1 = 10^{-1}$	deci	d	$dm = 10^{-1}$ m
$0.01 = 10^{-2}$	centi	c	$cm = 10^{-2}$ m
$0.001 = 10^{-3}$	milli	m	$mA - 10^{-3}$ A
$0.000,001 = 10^{-6}$	micro	μ	$\mu V = 10^{-6}$ V
$0.000,000,001 = 10^{-9}$	nano	n	$ns = 10^{-9}$ s
$0.000,000,000,001 = 10^{-12}$	pico	p	$pF = 10^{-12}$ F

Figure 11-4. Multiples and submultiples of numerical units.

there are both multipliers and dividers used in the system. The most common multipliers are k for kilo and M for mega. These letters represent 1,000 and 1,000,000, respectively. Resistors are commonly marked or discussed, using this system. It is much easier to write 12 k ohms (12 kilohms) than to show 12,000 ohms, in the technical literature. Also, it is less apt to be misread.

On the fractional end of the chart, the common values are m for 0.001 or 1/1000, μ for 0.000,001 or 1/1,000,000, and p for 0.000,000,000,001 or 1/1,000,000,000,000. These are often called milli, micro, and pico, respectively. There are other values, but these are the ones most commonly associated with electronic components. One way of helping to remember whether the value is a multiplier is that these units generally use a capital letter to represent the multiplier. The fractional values are shown with small letters. These units are used in the value description of the device. Typically, a reference would be for a 180 kilohm resistor, or a 180 k ohm resistor, instead of a one-hundred-eighty-thousand ohm resistor.

Safety.   Recent safeguards imposed upon set manufacturers include those related to fire safety. As mentioned earlier, resistors give off heat under normal operating conditions. If there is too much current moving through the resistor, it may operate beyond its safe rated value. When this occurs the resistor often literally *burns up*. This condition may possibly set the stage for a fire in the set.

Resistors used in critical circuits are identified by set manufacturers. These resistors are made so that they are flameproof and cannot cause a fire even if they are operated beyond the safe level. When changing one of these resistors, the technician should replace it with another flameproof resistor if the set is to be restored to its original operating and design condition.

Variable resistors.   Another kind of resistor, other than the fixed-value type previously discussed in this chapter, is the variable resistor. Pictures of variable resistors are shown in Fig. 11-5, along with their common schematic symbol. Most of these variable resistors have a fixed-value element connected to the two outside terminals. The middle terminal is connected to an arm which rotates inside the device. In this manner the center terminal is continuously variable over the entire fixed-resistance range. These variable resistors are available in a wide range of resistance values. They range from lows of 10 ohms or so to highs up in the megohm range. In all cases, the measured resistance from the arm to one terminal and from the arm to the other terminal is equal to the total fixed-value of resistance of the element. The two values will each change as the arm is moved on the element, but the total value of resistance remains the same.

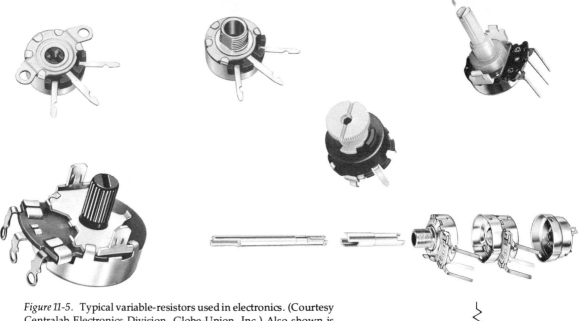

Figure 11-5. Typical variable-resistors used in electronics. (Courtesy Centralab Electronics Division, Globe-Union, Inc.) Also shown is schematic symbol.

Special resistors. There are three special kinds of resistors that operate quite differently from their normal counterparts. These are called *thermistors, varistors* and *light-dependent resistors.* Under normal operating conditions, resistors are not affected by heat, voltage, or light. A carbon resistor will usually increase in value when it gets too hot. The *thermistor,* on the other hand, is designed to reduce its ohmic value when heated. In a circuit, this resistor starts with an extremely high value. As current moves through the thermistor, it heats up. The material from which it is made exhibits what is called a negative-resistance characteristic. As it heats, its value reduces to a very few ohms and its circuit current increases. This kind of resistor is used to direct current movement. The circuit shown in Fig. 11-6 indicates how this works.

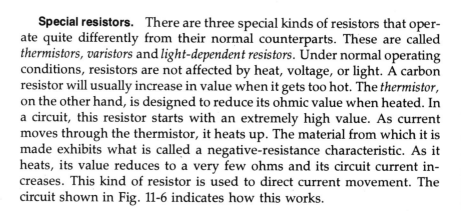

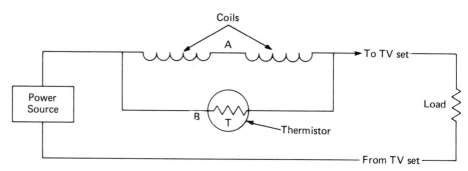

Figure 11-6. A circuit using a thermistor to direct electron flow.

Part A of the circuit consists of two coils of wire used to demagnetize a color picture-tube field. Part B is a thermistor. When the power is first applied to this circuit, the thermistor has a much higher resistance than the coils of wire. Current takes the path of least resistance. Therefore, the coils of wire, at A, will get most of the initial current and the thermistor, B, little current. As the thermistor decreases in value, the current shifts from path A to path B. The result is a bypassing of path A. The demagnetizing action is turned off and current is available to operate the rest of the set.

The varistor, or *voltage-dependent resistor* (VDR), is another of the special types of resistors. This resistor will change value with variations in voltage. As the applied voltage increases, the varistor will decrease in value. This effectively reduces the voltage developed across the varistor and tends to keep circuit voltage at a constant value.

The third device is the photo resistor, or *light-dependent resistor* (LDR). The LDR's resistance changes from a very low value, when there is a lot of light, to a very high value of resistance, in darkness. The LDR is used in automatic brightness-controlling circuits in the TV set. These three devices are pictured in Fig. 11-7, along with their schematic symbols.

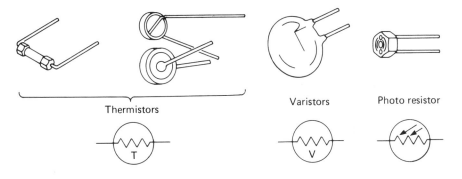

Thermistors   Varistors   Photo resistor

Figure 11-7. Special-resistor types.

Resistor applications.   The resistor is a functional device in any electronic unit. Its purpose is to oppose, or resist, the movement of electrons in the circuit. Using a resistance material, the resistor offers opposition to electron current. As a result of this opposition, a difference of voltage is developed between the two terminals of the resistor. This action is used to great advantage by circuit designers. If a device, such as a lamp, is designed to operate from a 12 volt source, it must have a specific amount of resistance in it. Otherwise, increasing the applied voltage will make the lamp brighter. Reducing the voltage measured at the connections to the lamp bulb will decrease the brightness of the lamp. The voltage is reduced by adding a resistance to the circuit as shown in Fig. 11-8. A voltage is developed across the added resistance. This voltage is taken from the source voltage. As a result, there is less voltage left to force current through the lamp.

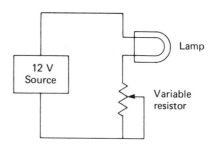

Figure 11-8. A circuit used to control current flow and voltage at the lamp terminals.

The term used for this voltage developed across the terminals of a resistor as a result of current flow is called *voltage drop*. If there is no current movement, then there is no voltage drop. In an incomplete circuit, where there is no way in which the current can flow, there is no voltage drop developed across resistor terminals. In such a circuit, the resistors act like pieces of wire with very little resistance. In complete circuits, the voltage drop developed across the resistances equals the applied voltage from the source. Keep in mind that, under normal conditions and in a complete circuit, all working electronic devices exhibit the characteristics of resistance. The exact amount of resistance will vary among different devices, but all of these various *loads* are resistive in nature.

11-2
Capacitors

The capacitor has but a single purpose. This is to store an electrical charge until it is needed and then to give up this charge to the circuit. The capacitor is made of two metallic surfaces and an insulating material placed between the plates. The insulating material is called the *dielectric*. It may be air, glass, mica, paper, or even oil, or any one of many other insulating materials. The unit of capacitance is the *farad*. This unit is too large to use conveniently in electronics work. Because of this, the terms *microfarad* (0.000,001 farad) and *picofarad* (0.000,000,000,001 farad) are used. The farad, is defined as being that amount of capacitance developed when a unit of electrical charge called a *coulomb* is stored in the capacitor by a one volt source. The coulomb is a very small unit. It is too small for normal use in electronics. Since the end result of this ratio is a very large unit, reference is normally to microfarads and picofarads.

Typical capacitors are shown in Fig. 11-9. There is a great deal of variety to the shape and form used for a capacitor, because few standards for them have, so far, developed. Efforts are being made to standardize some sizes, particularly for the replacement market, where several manufacturers are now producing capacitors that are interchangeable with respect to mounting holes and wire-lead placement. Size determines the total capacitive value. A general rule is that the larger the metallic plates,

Figure 11-9. Typical capacitors used in electronics. (Courtesy Mallory Distributor Products Co.)

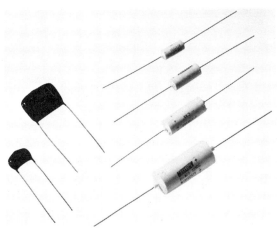

the larger the possible charge on them and the higher the value of capacitance. Actually, total capacitance depends upon the size of the metallic plates, the kind of dielectric insulation used, and the spacing between the plates. Capacitors are made in three basic forms. These are *fixed*, *variable*, and *adjustable*. Fixed types are shown in Fig. 11-9. Variable and adjustable types are shown in Fig. 11-10. Actually, both the variable

Figure 11-10. Typical variable and adjustable capacitors. (Courtesy, J. W. Miller Division, Bell Industries)

163

and the adjustable capacitor can vary. The difference lies in that the variable type usually has a shaft on it and has a control knob attached to the shaft, making it a customer operated control. Adjustable capacitors normally are "set" by the technician during manufacture or servicing of the set.

Capacitor types. There are a few basic capacitor types in use in home-entertainment equipment. These are generally classified as *electrolytic, paper* or *mylar,* and *ceramic* or *mica* capacitors. With the exception of the electrolytic type, all are essentially similar in operation. Paper or mylar types are made of long thin sheets of aluminum. The dielectric separating these two sheets is either paper or mylar plastic. The sandwich of aluminum sheets and dielectric is rolled into a tubular shape and wire leads are connected to each aluminum sheet. The capacitor is then dipped into wax or plastic to seal the unit. Its capacitance value is printed on the outside of this package.

Mica capacitors use mica as a dielectric. Ceramic capacitors are similar to the mica units except that the metallic surface is usually plated onto the ceramic. These units are capable of developing fairly high amounts of capacitance in a relatively small space.

Electrolytic capacitors are something special. These capacitors are polarity sensitive, whereas the others are not. If an electrolytic capacitor is connected into a circuit with its polarity reversed, it may explode and/or damage other circuit components. This capacitor is capable of developing relatively large amounts of capacitance in a small space. Some of these units are manufactured with two, three, or four capacitors contained in one metal-can housing.

Capacitor ratings. Capacitors are rated in two ways. One is according to capacitance value. The small disc-capacitor, rated in picofarads, has a numerical value printed on the side of the unit. Paper or mylar capacitors are normally made to cover a range of 0.5 to 0.0001 microfarads, and this value is printed on the container, also. Electrolytic capacitors are rated in microfarads, too. The value ranges are much larger, running from 1 to 2000, or more, microfarads, which values are also printed or stamped on the container.

The second method of rating the capacitor is according to its operating voltage. This rating specifies the maximum amount of voltage that the capacitor will tolerate before the dielectric insulation fails. Capacitors may be operated at voltages below their rated values, but never at voltages higher than the rating.

Capacitor applications. There are three major functions for which the capacitor may be used in a circuit. The capacitor may be used for coupling, bypassing, or as a part of a resonant circuit. A *coupling* capacitor is used to transfer signals from one circuit to another. This is done to eliminate the necessity of transfering operating voltages between circuits. A *bypass* capacitor is used to transfer a signal to common. In this manner unwanted signals are removed from operating voltage circuits. The capacitor works with an inductor in a *tuned circuit* in order to accept or reject a specific frequency or set of frequencies. This topic is discussed later in this chapter. Schematic symbols for capacitors are shown in Fig. 11-11. These are standard symbols. The value of capacitance as well as the voltage rating has to be provided in order to have complete information about the device.

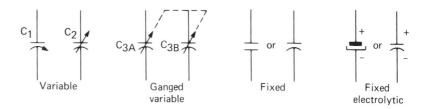

Figure 11-11. Standard schematic symbols for capacitors.

11-3
Inductors

An inductor in its simplest form is a coil of wire. The purpose of the inductor is to try to oppose any change in circuit current. The capacitor, when storing an electrical charge, tries to hold circuit voltage to a constant value. The inductor works in a similar manner to maintain circuit *current*. Here is how it works:

Any electric current moving in a wire will produce a magnetic field around the wire. If the current is constant, the size of the magnetic field is determined by the strength of the current. The magnetic field increases as the electron current increases. If the current changes, then the magnetic field also changes. A varying current produced in sine wave-form will produce a varying magnetic field in the same kind of form.

Scientists found that the magnetic field *when moving* is able to produce a voltage on another wire placed in the field. This occurs even though the second wire is not connected to the first wire. As long as the field expands or contracts the voltage is *induced* onto the second wire. When the field is expanding, the induced voltage has the opposite polarity of the voltage in the first circuit. When the field contracts, then the induced voltage has the same polarity as the first circuit. This principle is used in the electric generator. It is also used in a modified form in the electric motor.

It was found that the same properties of induction would occur in a single coil of wire. The magnetic field from one turn, or winding, will affect the winding next to it and any others within its magnetic field in the same way described in the previous paragraph. This form of induction is called *self-induction*. This means, simply, that the magnetic field tends to keep current constant in the coil of wire. As the current increases, the voltage induced on other wires, being of opposite polarity, tries to keep the current at the original level. When the current decreases, the induced voltage, now the same polarity as the source voltage, tends to keep the current up to the original level. Because of the manner in which it is designed, the coil of wire becomes a very effective inductor in a circuit.

The unit of inductance is the *henry.* The letter symbol for an inductor is L. Figure 11-12 shows several styles of inductors. Each has a specific use in a circuit. Some inductors are free-standing coils of wire. Others are wound on paper, ceramic, or metal forms. The core of some coils is made of ferrite. The ferrite increases the amount of inductance in the coil. Some ferrite-core inductors are adjustable while others are fixed value. The amount of inductance possible in any inductor depends upon several interrelated factors. These include wire size, diameter of the coil, length of the coil, spacing between turns of wire on the coil, and the kind of core material used in the inductor.

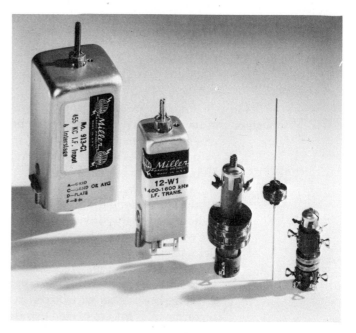

Figure 11-12. Typical inductors used in electronics. (Courtesy J. W. Miller Division, Bell Industries)

Inductors are rated in henries or sub-units such as *millihenries* or *microhenries.* They also have a voltage rating in some instances. For the most part this voltage rating applies to the insulation of the inductor. Too high an applied voltage will destroy the insulation of the inductor and make it ineffective in the circuit.

Circuit applications. Inductors are seldom used alone in a circuit. Most of the time they are found in combination with resistors or capacitors. Examples of circuits using these combinations are the delaying circuit, the wave-shaping circuit, and the filtering circuit. The inductor is used with a capacitor in circuits that are frequency selective. Each type of circuit, and specific applications of inductors, is presented in later chapters.

11-4
Transformers

The transformer is an adaptation of the inductor. Physically, the transformer consists of two or more coils of wire (inductors) placed so that each coil of wire is close to the others. Energy is transferred from one coil to the others by means of inductance. One coil is identified as the *primary*, or input, winding. The other winding, or windings, are called secondary windings. Most transformers have a single primary winding and one or more secondary windings. Photographs of transformers found in home-entertainment devices are shown in Fig. 11-3.

Transformers are designed to work within various frequency ranges. These include power, audio, and radio or intermediate frequencies.

Figure 11-13. Typical transformers used in electronics. (Courtesy Triad-Utrad.)

Power transformers normally use an iron core material and are often physically larger than other types. Audio-frequency transformers also use iron cores. They're usually smaller than power transformers because they handle less power. RF and IF transformers often have ferrite cores and are quite small by comparison to other types.

Power transformers transfer power from one circuit to another. The rule of "power out equals power in" applies to these units. The voltage at the secondary winding of the transformer may be higher or lower than the input voltage. This voltage is part of the power formula. If the voltage is *stepped up*, then the available current in the secondary will be less. For example: a power transformer has a secondary power rating of 600 watts. The voltage developed in the secondary is 600 volts. The load is such that 1 ampere of current flows. Using the power formula,

$$\text{Watts} = \text{Current} \times \text{voltage}$$

The power is equal to 600 × 1, or 600 watts. If 600 watts of power is working in the secondary, then 600 watts of power is working in the primary. The primary current would be figured by using Watt's law. The source voltage is 120 volts.

$$\text{Current} = \frac{600}{120}, \quad \text{or 5 amperes}$$

In another case, with a similar transformer, the secondary voltage is stepped *down* to 24 volts. Now, current = 600/24 and there are 25 amperes of current available to do whatever work is required in the circuit. The voltage-ampere ratio of the transformer may vary as long as the total power rating of the transformer is not exceeded. The result of an operation that is over the power rating is usually smoke, hot tar, and a ruined transformer.

When a multiple secondary-power transformer is used, the power ratings of each secondary are added to determine the total secondary power. Here, too, the total secondary power cannot exceed the primary power.

Audio, RF, and *IF* transformers transfer electronic signals between stages of the device in which they are used. These transformers are not designed to handle power but, instead, are frequency sensitive. These transformers are designed to permit a specific group of frequencies to be transferred from the primary to the secondary, while all other frequencies are reduced to such a low amplitude that they are, in effect, stopped. Schematic symbols for various transformers are shown in Fig. 11-14.

**11-5
Resonance**   When something is in resonance it vibrates at a specific frequency. This is an important factor in electronics. Many resonant circuits are

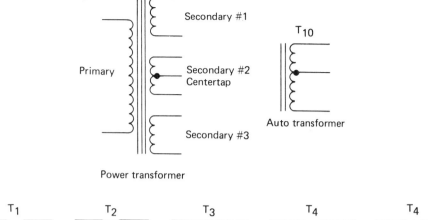

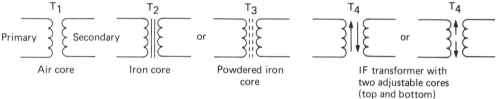

Figure 11-14. Schematic symbols for transformers.

required in order to have a functioning radio or TV set. These circuits are often called *tuned circuits*. A *resonant circuit* is made up of a capacitor and an inductor. These circuits function with varying voltages or signals. Since this is the case, the circuit's resistance is not considered. Instead, the circuit's reactance becomes an important factor. Reactance is the opposition to electron movement in a circuit with constantly changing values. Both capacitive reactance and inductive reactance are involved in a resonant circuit. When these two factors are equal, then the circuit is resonant. The frequency of resonance is determined by the sizes of the capacitor and inductor. In general, the resonant frequency is low when they are large and high when they are small. Changing any component in the circuit will change the resonant frequency.

There are two types of resonant circuits in general use. These are called *series resonant* and *parallel resonant*. Each has specific characteristics. They are illustrated in Fig. 11-15 along with a listing of their main characteristics. Use is determined by the engineer when the set is designed. Selection is based upon whether it is desired to transfer maximum signal current, with the series-resonant circuit, or to transfer maximum signal voltage, with the parallel-resonant circuit.

Many tuned circuits are *variable*. If either the inductor or the capacitor is a variable unit, then the resonant frequency is also variable. This principle is used in tuning-devices for radio and TV sets. Tuning is accomplished in a radio by using a fixed inductor and a variable capacitor. Other devices use a fixed capacitor and a variable inductor to select a

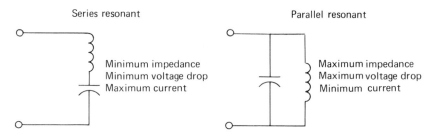

Minimum impedance
Minimum voltage drop
Maximum current

Maximum impedance
Maximum voltage drop
Minimum current

Figure 11-15. Resonant circuits and their characteristics.

specific frequency. Both ways are good. Selection is usually dependant upon the specific physical properties and cost factors in the unit being produced. Some more recently produced devices are using a semiconductor called a *varactor diode* instead of using a variable capacitor. This unit, which is simply a semiconductor whose capacitance may be changed by applying a voltage to it, is discussed in detail in Chap. 13. Changing the applied voltage changes the capacitance of the device. When the *varicap* (variable capacitance) is placed in a tuned circuit, it acts just as a capacitor acts.

Two other terms important to those who wish to understand about tuned circuits are *bandpass* and *Q*. Bandpass refers to the specific group of frequencies that are accepted and passed through the tuned circuit. Normally, this group of frequencies is identified by one of the major frequencies involved. An IF transformer used in an AM radio has a resonant frequency of 455 kilohertz. The modulation on this frequency extends 5 kilohertz above and 5 kilohertz below 455 kilohertz. The bandwidth of an IF transformer used in an AM radio has to be 10 kHz in order to pass all necessary frequencies.

The bandwidth and selection capabilities of a circuit determine the quality of the tuned circuit. This is commonly called the *Q* of the circuit. As the selectivity of the circuit and the bandwidth narrows, the circuit *Q* increases. A tuned circuit that has a wide bandwidth is said to be a low *Q* circuit, while one exhibiting the opposite characteristic is identified as a high *Q* circuit. Circuit *Q* is design factor and is presented in order to broaden the general knowledge of the technician at this time.

11-6 Magnetism

Most of us are aware of the magnet. We know that certain metals are able to exhibit magnetic properties. These properties include the strength of the magnetic field and the existence of magnetic poles. We also remember that opposite poles attract each other and that similar poles repel each other. These same principles apply to the electromagnet. A coil of wire wound around an iron form exhibits the characteristics of a magnet when an electric current flows through the coil. Stopping the current flow shuts off the electromagnet.

This principle is used in several ways in home-entertainment devices. One way is in the motor used to operate tape players or phonograph turntables. Establishment of a variable magnetic field in a motor will cause the motor to rotate. The magnetic field in this case varies with the power line frequency. This is used to provide a constant-speed motor-drive system.

Another use for electromagnetism is in the device called the *solenoid*. Physically, the solenoid is a coil of wire wound in the form of a helix. A piece of iron core-material is placed in the hollow center of the coil. This is usually attached by a spring to a metal mounting-plate. When the coil is energized the core moves out of the form. When the core is attached to a mechanical linkage, work is performed. An application of this is in the tape-head position-changing mechanism used with an 8-track tape unit. This is illustrated in Fig. 11-16. When the circuit is completed by closing the push switch or when the sensing switch is operated, current moves through the coil of wire. This makes the plunger move which, in turn, makes the change arm operate the ratchet plate. The ratchet plate is used to position the tape play-head.

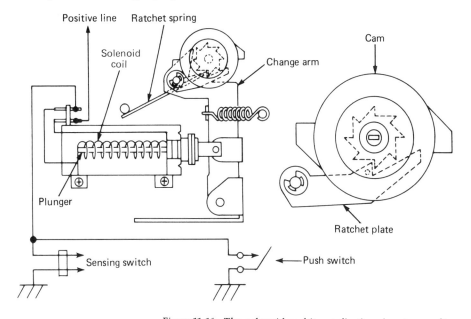

Figure 11-16. The solenoid and its applications in a tape unit.

11-7
Circuit
Protection Devices

Most electronic devices have some circuit protection. One kind is the *overcurrent* protective device. This is an intentionally weakened part of the circuit. It is designed to fail when an overcurrent condition exists. There are two basic types of overcurrent circuit protectors. These are the *fuse* and the *circuit breaker*. Both are shown in Fig. 11-17. Basically, the fuse is a piece of wirelike material which is sensitive to temperature. As the temperature of the fuse link rises the material melts. Fuses are rated in three ways. These are according to current, voltage, and melting characteristic.

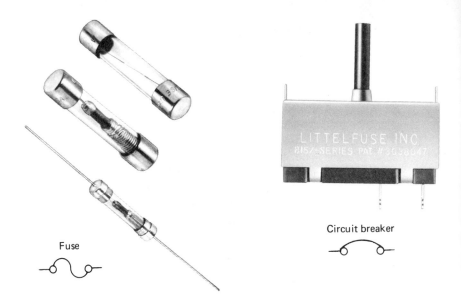

Fuse

Circuit breaker

Figure 11-17. Fuses, and a circuit breaker with their schematic symbols. (Courtesy Littelfuse, Inc.)

All fuses are rated by their *current* handling ability. Cartridge fuses used in electronic devices may be designed to handle a low current of only 0.002 ampere. The high end of the range is 50 amperes. There are about 60 standard values of fuses between these two extremes.

The second rating on a fuse refers to its safe operating *voltage.* The standard test for this rating is a circuit carrying 10,000 amperes of current at operating voltages of 32, 125, and 250. These three values handle most conditions used in automobile and electronic applications. This voltage rating is not an absolute value. The fuse may be used safely in circuits with higher operating voltages if a short-circuit condition will not produce high arcing in the fuse.

The third rating for a fuse is according to its *blowing* time. Fuses are rated as fast acting, normal acting, and "slo-blo" acting. This refers to the time required to cause the fuse to melt under overcurrent conditions. The "slo-blo" fuse is used in circuits that require high currents as the device is turned on but much lower current after the initial surge. These fuses will handle a 100% overload (2 × rated value) for a period of about 10 seconds before they melt.

In addition to the above three characteristics, *size* is another way by which fuses are identified. There are two standards for fuse size ratings in use today. One uses 1 AG, 3 AG, etc. The other uses AGA, AGC, etc. Both refer to the physical size of the fuse. Most cartridge fuses in car or electronic use are ¼ inch in diameter and vary from ⅝ inch to 1-⁷/₁₆ inches in length. The *AG* rating tells the size of the fuse.

172

The other overcurrent protective device is the *circuit breaker*. When used in home-entertainment devices, the circuit breaker is a manual-reset device. Ratings for circuit breakers range from 0.65 to over 4.14 amperes. The breaker is made with a bimetal strip and a set of contacts. When the bimetal strip heats, it bends moving one contact away from the other contact. This interrupts the circuit current and shuts off the device. A spring mechanism holds the contacts open until the manual-reset button is pushed. The advantage of the circuit breaker is that it is reuseable, while the fuse is a one-time device.

Semiconductor diode.   Another kind of circuit protective device is the semiconductor diode. The diode is not as obvious a protector as the fuse or circuit breaker. It is used to protect against voltage surges or reverse-polarity power connections. The semiconductor diode operates like a voltage-sensitive switch. It serves two purposes. The diode is *on* when the internal resistance of the diode is low or *off* when the internal resistance is extremely high. The condition depends entirely upon the amount of applied voltage. Also, if the polarity of the voltage is correct, then the diode acts as if it were *on*. If the polarity were accidentally reversed, then the diode would be *off*, the current would be little, if any, and the circuit would be protected.

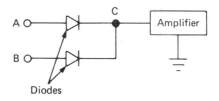

Figure 11-18. A diode steering-circuit.

This principle is used to *steer* electrons or signals in a circuit. Refer to Fig. 11-18. In this circuit signals may be applied to points A or B. When a signal is at A, and its polarity is proper, the signal will pass through the diode and appear at C, or the input of the amplifier. If this is changed and the signal is at B, then the other diode is on, the first one is off, and the B signal becomes the input to the amplifier. This type of circuit is used to steer signals in a set so that they may be processed.

In still another instance, the diode may be used to protect the circuit against over-voltage conditions. In this type of circuit, the diode is connected across the power-source output terminals. If the voltage rises to too high a value, the diode acts as a low resistance path and effectively reduces the higher voltage to the desired value. When the operating voltage is at or below the desired value, the diode acts as if it is not in the circuit.

SUMMARY
Twelve basic concepts and components are identified in this chapter. The components are resistors, capacitors, and inductors. Three basic concepts are resistance, voltage, and current. The circuit configurations are series, parallel, and combination series-parallel. Amplifiers, oscillators, and rectifiers are three kinds of electronic circuits. These twelve items are fundamental to a knowledge of electronics. They are considered required knowledge for persons employed as repair technicians. Both the knowledge of how these items function and how they work in practical circuits is required if one is to be successful in the repair field.

There are two extremes in materials used in electronics. At one end is any material which does not allow electron movement. This material is classified as an insulator. The opposite of this is any material which permits easy electron movement. This is a conductor. Between these two extremes is a large group of materials called semiconductors.

The most common device in use in electronics is the resistor. Its purpose is to slow electron movement in the circuit. A result of this action is that a difference in voltage develops across the resistor in a circuit. This voltage drop is used to provide different voltage levels to circuits. The resistor is rated in an ohm value, and it also has a power rating. Both are considered when selection of a resistor for a circuit is made. Resistors are manufactured as either fixed or variable units. Special resistors include voltage-dependent, light-sensitive and heat-sensitive types. These three special resistors are used in circuits to control electron flow.

The capacitor has the ability to store an electric charge. It is stored until the charge is needed in the circuit, at which time the capacitor releases its charge into the circuit. The capacitor is made of two metallic plates separated by an insulating dielectric material. The capacitor is rated according to both its charge value (farads) and its voltage (dielectric breakdown) value. Capacitors are made in both fixed and variable units.

Inductors are essentially coils of wire. A current passing through a wire produces a magnetic field around the wire. As this field expands and contracts, it may produce a voltage on another wire in the same field. This process is called induction. A voltage is induced on this second wire every time the field moves over the second wire. The voltage will either add or subtract from the circuit voltage, depending upon the direction of movement of the magnetic field. This voltage tends to maintain the circuit current. Inductance in a circuit tries to keep electron current constant.

Inductance may be increased by use of a core material in the inductor. Core materials used are ferrite and soft iron. High-frequency inductors use a ferrite core while the low-frequency inductors use soft iron for core material. Inductors are made as either fixed or variable. Variable inductors have a moveable core.

The transformer is a form of inductor. It usually has a primary winding and one or more secondary windings. Voltages may either be increased or decreased in the transformer. This depends upon the ratio of wire windings between the primary and the secondary. The power total is always constant. *Power out* still has to equal *power in*.

A circuit containing resistance, inductance, and capacitance will resonate at a specific fequency. These tuned circuits are used to accept or reject specific frequencies in electronics. Two basic circuits are the series-resonant and the parallel-resonant circuits. Each has its own characteristics in a circuit.

Magnetism is used in electronics work in motors for operating turntables and tape units. It is also used as a control unit in the solenoid. The solenoid is an electromagnet with a moveable core. The core is connected to a linkage or mechanism which operates as the core is moved. One common use is to index, or move, the head in an eight-track recorder.

Fuses and circuit breakers are used to protect electrical circuits. These are both heat-sensitive devices. Too much heat will melt the fuse or cause the breaker to open. A product of electron current is heat. Too much current causes the fuse link to heat to the melting point. When it melts, it protects the circuit by stopping current movement.

Another circuit protective device is the semiconductor diode. It has the ability to act as a switch in the circuit. This action is dependent upon the polarity of the voltage across the diode. This action is used to operate circuits as well as to protect them.

QUESTIONS

1. What is a conductor? an insulator? a semiconductor?

2. Why are power ratings important to a resistor?

3. What is the resistance rating of each of these resistors?
   (a) red, yellow, orange, silver
   (b) red, red, red, gold
   (c) orange, white, orange
   (d) brown, black, red, silver
   (e) violet, green, black, gold

4. Write the color bands for these resistors.
   (a) 12 kilohm ±5%
   (b) 100 kilohm ±20%
   (c) 2.2 megohm ±10%
   (d) 27 ohm ±5%
   (e) 4.7 kilohm ±10%

5. Why is a 2-watt rated resistor larger than a ¼-watt rated resistor?

6. Draw the schematic symbols for the following:
   (a) fixed resistor
   (b) variable resistor
   (c) fuse

(d) variable capacitor
(e) circuit breaker
(f) fixed capacitor
(g) thermistor
(h) inductor
(i) transformer with one primary and three secondary windings

7. Describe thermistor action.

8. Describe varistor action.

9. What is the purpose of a capacitor in a set?

10. What is the purpose of an inductor in a set?

11. What is meant by the term *resonance?*

12. How are fuses rated?

Circuits and Circuit Analysis

12

Success in repair of electronic devices is based upon many factors. Among these are the knowledge of how the device is developed on a block basis, the ability of the technician to localize the problem to a specific block or area, and then the competence to locate the specific component that has failed. Good technicians are able to locate defective parts by analyzing the circuit. Test equipment is employed to measure operating voltages and to locate points at which the signals become abnormal. Once the point of abnormality is found in the set, the technician uses additional test equipment to test and locate a specific bad part. Knowing the basic circuits—either series or parallel—and the laws which apply to each circuit, the technician is able to make a rapid, correct repair. This chapter presents material covering the basic circuits, and discusses the relationship of voltage, current, and resistance in these circuits.

A short review of some basic terms is in order at this time. These terms were introduced in earlier chapters. They will now be used to help explain how each circuit functions. When this is understood, then the understanding of electronics and related circuit analysis is a comparatively easy matter.

**12-1
Definitions**

Voltage. Electron movement in any circuit depends upon the force available to make the electrons do work. This force is called the *electromotive force* (EMF) because it is the force that makes the electrons move. The EMF is actually an electrical pressure in the system. As the pressure increases, more electrons will move in the circuit. The EMF is measured in *volts*. Voltage is actually based upon a difference in electrical pressure between two points in an electrical circuit. Such a statement as "there is 100 volts at that point" is actually an incorrect description. What is meant is that "there is a difference of 100 volts between that point and a common reference point in that circuit." Technicians understand the intended meaning and the use of the shortened statement. In other words, all voltage measurements are taken using a point of reference in a circuit. This point is called a *common* or *ground*.

Polarity. Voltages found in electronic equipment may be either positive or negative. These polarities are assumed to be with respect to the circuit common point. The polarity of the voltage depends upon where the power source is connected to the circuit. Refer to Fig. 12-1.

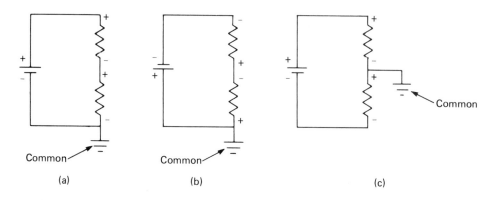

(a) (b) (c)

Figure 12-1. One circuit may have a variety of polarities, depending upon where the common point is located.

One circuit is shown in three different arrangements in this illustration. Figure 12-1(a) shows a power source and two resistors. The common point in this circuit is at the negative terminal of the circuit and power source. All voltages measured in this circuit are positive *with respect to common*. Figure 12-1(b) is the same circuit except that the polarity of the power source has been reversed. All voltages are negative *with respect to common* in this circuit. The third circuit, shown in Fig. 12-1(c), uses the same component parts. The only difference is in the point of the circuit identified as common. When a circuit with this arrangement is used, it is possible to have both a positive voltage *with respect to common* and a negative voltage *with respect to common*. What arrangement is used is decided by the engineer when the set is designed. It is important to remember that the polarity of the voltage depends upon two factors: the polarity of the source voltage, and where the common point is located in the circuit. All measurements are made with reference to the circuit common, whether this is stated or not.

Current. Current is defined as the movement of electrons in a circuit. Actually, there are two electrical-current descriptions in books relating to electricity. Early scientists believed that electrical current moved from a positive area to a less positive, or negative, area. This is called *conventional* current movement. Studies in later years have been based upon the movement of the electron. The electron is a negatively charged body. It moves from a mass of negative charges (other electrons) to an area containing fewer negative charges. Therefore, we now consider the electron movement to be from a negative field to a less negative, or positive, field. This book considers electron movement from negative to positive as the basis for understanding circuits and how these circuits function. This movement takes place regardless of the polarity of the voltage source or the placement of the common point in the circuit.

Electron current can be thought of as the quantity of electrons moving past a given point in one second. This current is measured in *amperes*. When quantities of electrons less than an ampere are measured, they are measured in *milliamps* (1/1,000) and *microamps* (1/1,000,000). All three of these terms are used in electronics.

Ac and dc. The easiest way to understand electrical circuits is to first consider one-way current movement. Figure 12-2(a) illustrates this. The arrows indicate the direction of electron flow. It is from negative to positive, and flows *through the circuit*. Reversing the polarity of the applied voltage, as illustrated in Fig. 12-2(b), will make the electrons flow in the opposite direction. These are both examples of a *direct current* (dc) circuit. The current movement is in only one direction. It never reverses

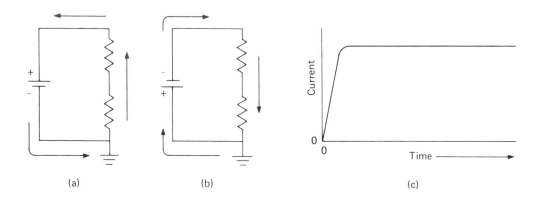

(a)   (b)   (c)

Figure 12-2. Electron flow in the circuit is from negative to positive. It remains constant in the dc circuit.

or moves in any other way. The amount of current depends upon two other factors. These are the amount of applied voltage and the amount of resistance in the circuit. A graph of the current movement in this circuit is shown in Fig. 12-2(c). At the first moment in time, the circuit is turned off and there is no electron flow. When the circuit is turned on, the current rises to a value determined by the applied voltage and the circuit resistance, and remains at a constant level after the initial turn-on time. This type of circuit is referred to as a dc circuit. There is another type of circuit in use as well as the dc circuit. It is called the ac circuit.

Ac is the abbreviation for *alternating current.* To alternate, here, means to constantly reverse direction. Ac constantly changes direction at a specific rate. This rate is called the *frequency* of the current. When the ac voltage changes, it is possible to change the polarity of the applied voltage. The polarity determines the direction of current movement. This is shown in Fig. 12-3(a). The solid lines represent the situation where the negative is connected to the circuit common. The reversal of the polarity of the applied voltage causes the electrons to slow down and actually stop moving. When the applied voltage polarity reverses, the electron flow also reverses, as illustrated by the dashed lines in Fig. 12-3(a). The waveform of this current is shown in Fig. 12-3(b). It varies in step with the applied voltage. This is representative of ac voltage in any circuit.

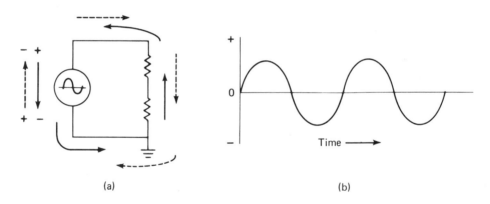

(a) (b)

Figure 12-3. Electron flow in an ac circuit constantly changes direction and value.

Some circuits may have only a dc voltage or an ac voltage. Other circuits may have both voltages operating at the same time in them. This is shown in Fig. 12-4. A dc voltage, of 10 volts, and an ac voltage, varying from +2 to −2 volts, are applied to the same circuit. When two voltages of any polarity or shape are applied to one circuit at the same time they are added together. The result of this action is shown in Fig. 12-4(b). It is a voltage varying from the 10 volt dc level by +2 volts and −2 volts. This type of circuit will be discussed further in a later chapter.

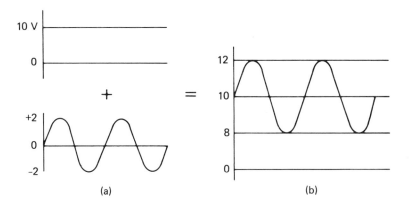

Figure 12-4. Two applied voltages add to each other to form a new applied voltage.

Interactions of voltage, resistance, and current. Voltage, current, and resistance are independent factors in any electronic circuit. While they are independent, they must work together if the circuit is to function. The interrelationship of these three factors is identified in *Ohm's law*. This law states that one ampere of current moves in a circuit containing one ohm of resistance when one volt is applied to the circuit. This relationship is shown as a formula:

$$E \text{ (voltage)} = I \text{ (current)} \times R \text{ (resistance)}$$

The interrelationship is illustrated in Fig. 12-5. The Ohm's law triangle works just like the power triangle described in Chap. 8. If one wishes to find any one of the three factors and the other two are known, it is easy to solve, after first substituting actual values for the letters. For example, how much current is in a circuit having 100 ohms of resistance and an applied voltage of 10 volts? Using the Ohm's law triangle and substituting known values, $I = E/R$, or $I = 10/100$. The answer is 0.1 ampere. This formula is used to find voltage, current, or resistance in any circuit.

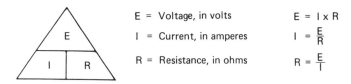

E = Voltage, in volts $E = I \times R$

I = Current, in amperes $I = \frac{E}{R}$

R = Resistance, in ohms $R = \frac{E}{I}$

Figure 12-5. Ohm's law triangle and the variations of Ohm's law.

12-2
Circuit Analysis   Each electrical circuit is composed of three basic parts. These are the *power source*, the *load*, and the lines or *wires* connecting the load to the power source. This is illustrated in Fig. 12-6. The load is in reality the

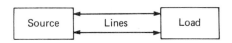

Figure 12-6. A basic electrical circuit, consisting of a source, a load, and two lines.

device doing the work in the circuit. Radios, tape units, and TV sets are typical home-entertainment loads. Electrical energy provided by the source performs work in the load. It is necessary to have some connection between the load and the power source. This is accomplished by use of wires. Removal of any one of the three components that go to make up the electrical circuit results in an incomplete circuit and no work can be performed.

When the basic circuit is identified then it is analyzed in order to determine how the electrons are working. There are three basic circuits used in electronics. These are called series, parallel, and combination series-parallel. The relationship of the placement of the load (or loads) with respect to the power source is used to analyze any circuit. If one assumes that the simpler circuits have each only one source of power, then a look at the loads and how they are connected to the source will determine whether the circuit is either series, parallel, or a combination of series and parallel components.

Parallel circuit. The *parallel* circuit is one that has two or more loads directly connected to the power source. A parallel circuit is shown in Fig. 12-7. The circuits shown in the drawings are the same. Each has one source and three loads. The voltage at the power source causes electrons to move through each of the loads. Figure 12-7(a) shows three sets of wires, one for each of the loads. Figure 12-7(b) gives the same information in a manner which is more familiar to the electronic technician. Each of the loads is connected directly to the source in both of these drawings.

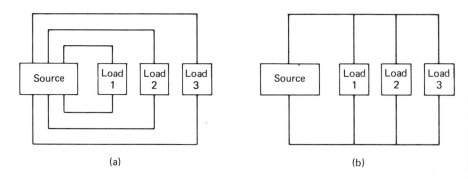

(a)   (b)

Figure 12-7. The parallel circuit. The right-hand version is a simpler schematic form.

Certain simple conclusions about parallel circuits may be drawn. The parallel circuit has the same voltage at each load, and this voltage is the same as the source voltage. Ohm determined that the electron movement through a load depends upon the applied voltage and the resistance of the load. In the parallel circuit, since the applied voltage is the same for all loads, the electron movement depends upon the resistance of each of the loads. This electron current may be measured as it moves through each load. In this circuit, current through each load is added in order to find the total current in the circuit. Ohm also found that, as more loads are added in a parallel circuit, the current increases. This indicated to him that the opposition to this current must decrease as more loads are included in the circuit. There are three fundamental rules for parallel circuit. They are:

1. In the parallel circuit, the voltage at each load equals the source voltage.
2. In the parallel circuit, current through each load depends upon the resistance of that particular load. These individual currents add up to equal the total current in the circuit.
3. In the parallel circuit, as loads are added to the circuit, there is more current moving and less total opposition.

The rules for parallel circuits may also be shown in mathematical form. These are:

1. $E_{applied} = E_{load\ 1} = E_{load\ 2} = E_{load\ 3}$

2. $I_{total} = I_{load\ 1} + I_{load\ 2} + I_{load\ 3}$

3. $\dfrac{1}{R_{total}} = \dfrac{1}{R_{load\ 1}} + \dfrac{1}{R_{load\ 2}} + \dfrac{1}{R_{load\ 3}}$

A common place to find a parallel circuit is in the home. In the home branch circuits are connected from the fuse panel to the various rooms. These circuits are wired in parallel so that the voltage is constant at each outlet. As stated in rule 2, the current will vary depending upon the size of the load connected to the source. Let's look at the wiring found in a typical kitchen and at the current used as appliances are turned on. The values used in this example are not accurate for all appliances and are used for discussion purposes only. A refrigerator uses 3 amperes of current. The coffee pot uses 2.5 amps. A radio may use 1 amp. Lights may use another 2 amps. The toaster uses 8 amps. When all of these devices are turned on, the total current use is 16.5 amps. The current used will vary depending upon the number of loads. Each of the loads has a specific resistance. As more loads are connected, the *total* resistance in the circuit goes down and the current increases. In some instances, when the resistance of a load is too low, too much current is used. If a fuse or circuit breaker is in the circuit, this protective device should open and turn off the circuit. Excessive current flow also occurs when a short circuit develops. The short circuit is actually a very low resistance connection. Low resistance causes large amounts of current to flow. Too much current can cause a fire if the circuit is not protected.

Power in the parallel circuit is determined simply by adding the individual power consumption figures. Using the kitchen just described, as an example, we can determine the amount of power used:

Refrigerator	$P = E \times I$ = 120 V ×	3 A =	360 W
Coffee pot	= 120 V × 2.5 A =		300 W
Radio	= 120 V × 1 A =		120 W
Lights	= 120 V × 2 A =		240 W
Toaster	= 120 V × 8 A =		960 W
Total power consumption		=	1,980W

When all of the circuits are turned on, almost 2,000 watts of power are used. Analysis of this circuit shows that each load in the parallel wired circuit does indeed meet the rules. All of the loads have 120 volts applied to them. The individual currents add up to total 16.5 amperes. The power formula, $E = I \times R$, agrees with the addition of the individual power ratings. The circuit resistance could be proven to be about 7.27 ohms.

The author uses two methods when analyzing circuit values. One is to approximate the values. Most values found in components and circuits are not exact. All have tolerances. Some may vary by as much as 20% from the marked value and still be acceptable. Because of this permitted deviation, the measured values in the circuit often will not agree with the calculated values. The author uses the formulas and rounds off the values in order to find the approximate answer. If the measured value is fairly close to the mathematically determined value, then the circuit is probably working.

When using the formulas to find circuit values, it may be easier to draw small Ohm's law triangles next to each load and the source. This method of finding voltage, current, and resistance values is very simple. For example: two loads are wired in parallel and connected to a 120-volt source. One load is rated at 6 amps and the other load is rated at 2 amps. Look at Fig. 12-8. There are Ohm's law triangles placed next to each part of the circuit. The known values written in each triangle, in the proper section. We know that each load operates from a 120 volt source. These may be inserted in the E section. Both currents were given. Fill in these also. Now it becomes easy to find the unknown. The resistance value for load 1 is 20 ohms. The resistance for load 2 is 60 ohms. Total resistance, figured at the source, is based upon total applied voltage and total current. It is 15 ohms. This method may be used with any type of circuit *as long as the rules for that circuit are followed.*

184

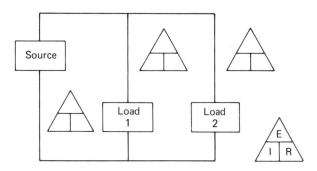

Figure 12-8. Use of Ohm's law triangle to find values in a parallel circuit.

The second method for analyzing circuit values is to proportion the values in a circuit. This works for any circuit. If one keeps the rules for circuit analysis in mind, then it is not difficult to proportion these values. Use the circuit presented in Fig. 12-8 as an example. Usually, the applied voltage and the resistance values are easiest to learn. Load 1 has a resistance of 60 ohms. Load 2 has a resistance of 20 ohms. The ratio of resistances is 60 to 20, or 3 to 1. Ohm's law shows that load 1 had a smaller current when compared to the lower resistance of load 2. Total current in this circuit is 8 amps. Load 1 has ¼ of the total current (1 of 4 parts), or 2 amps. Load 2 has ¾ (3 of 4 parts), or 6 amps. This alternate method is used when approximate values are satisfactory.

Series circuit. The second basic circuit arrangement is the series circuit. This differs from the parallel circuit in that all loads are connected to each other and the source in a different manner. A series circuit is shown in Fig. 12-9. The key difference between the parallel and the series circuit is that the series circuit has only one path for electron movement. All

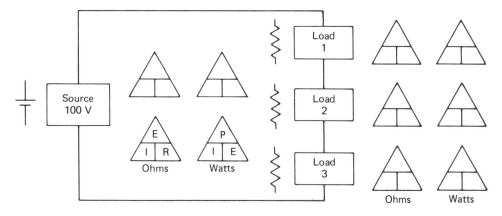

Figure 12-9. Use of Ohm's law and power triangles to find values in a series circuit.

electron current moves from the source and, in turn, through each of the loads, eventually returning to the source. In the series circuit, current movement is the same in each part of the circuit. Another term must be introduced at this point: *voltage drop*. The opposition to electron movement in a resistance will produce a difference of voltage across the resistance. This difference in voltage value is called *voltage drop*. This voltage drop has been defined as the work done as the electrical charges move through a resistance.

In the series circuit, a voltage drop develops across each of the loads. These individual voltage drops add up to equal the source voltage. When Ohm was working with electric circuits, he found that when the resistance is increased in the series circuit fewer electrons move in the circuit. When the resistance is decreased more current is measured. Applying this concept to the series circuit, we can state that, if more loads are placed in the circuit, there will be less current moving. Summarizing the discussion on the series circuit, we can state three rules for this kind of circuit:

1. In the series circuit, electron current is the same for all loads.
2. In the series circuit, the voltage drops across each load add up to equal the source voltage.
3. In the series circuit, as more loads are added, resistance increases.

These rules may also be shown in mathematical form. They are:

1. $I_{total} = I_{load\ 1} = I_{load\ 2} = I_{load\ 3}$

2. $E_{total} = E_{load\ 1} + E_{load\ 2} + E_{load\ 3}$

3. $R_{total} = R_{load\ 1} + R_{load\ 2} + R_{load\ 3}$

Series circuits are not often used in home wiring. They are commonly found in the circuits within consumer electronic products.

Circuit analysis for the series circuit is accomplished by either of the two methods described for parallel circuits. The term *voltage drop* was mentioned above. Another term used quite frequently in describing electronic circuits is *voltage divider*. A voltage divider is often a *series* of loads, or resistors. This is illustrated in Fig. 12-9. The applied voltage is divided across each of the three loads. The amount of voltage present across each individual load depends upon the resistance of each load. Keep in mind that the higher resistance in a series circuit (or voltage divider) will have a larger voltage drop. Kirchhoff's laws for a *closed loop* apply in this case.

Kirchhoff's circuit laws. Two other laws often used in circuit analysis are called *Kirchhoff's laws*. They are based upon Ohm's laws but go a step beyond the work done by Ohm. One of Kirchhoff's laws relates to the current at a junction in a circuit. It states that the current entering a junction is equal to the current leaving the same junction. In other words, current cannot accumulate in a circuit. This is illustrated in Fig. 12-10(a). The 3 amps of current entering the junction at "A" is equal to the sum of the currents through each of the loads.

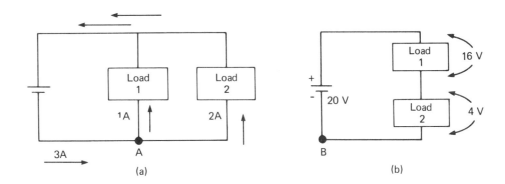

Figure 12-10. Application of Kirchhoff's laws for circuit analysis.

The other law relates to voltage distribution in a circuit. This is shown in Fig. 12-10(b). Kirchhoff said that the sum of the voltage drops developed in any closed-loop circuit is equal to zero. Start at point "B" and go clockwise around the circuit. Add each voltage drop including the source. Observe the polarity as shown in the illustration.

$$-20 \text{ (source)} + 16 \text{ (load 1)} + 4 \text{ (load 2)} = 0$$

A common way of stating this law, when only one source is used, is that the voltage drops in the load equal the source voltage. Individual voltage drops will vary depending upon the actual resistance involved, but the rule will always work. There cannot be any voltage left over or a shortage of voltage in any loop circuit.

Finding answers for circuit values is very easy if either the proportion rules or the mathematical rules are used. Some information has to be known in order to find these answers. Most of the time, the applied voltage and the individual load-resistance values are known. In Fig. 12-9 there is an applied voltage of 100 volts. The load resistances are 150, 250, and 600 ohms. Write these values in the correct box in each Ohm's law triangle. There still is not enough information to find answers. Two of the values must be known in order to solve for the third value. A convenient way to handle a problem of this type is to find the total values first. The rules for series circuits are used. Total voltage is equal to the applied voltage of 100 volts (rule 2). Total resistance is the sum of all resistance values (rule 3). Total resistance in this circuit is 1,000 ohms. The total current is found by using Ohm's law. The answer is 0.1 amps.

Rule 1 for series circuits shows that current is equal throughout the circuit. This means that total current is the same as the current in each load. The total current value of 0.1 amp is placed in each of the Ohm's law triangles. This leaves only the individual voltage drops as the unknown. Multiplying 0.1 by each resistance value gives the three individual voltage drops. They are 15, 25, and 60 volts. These drops add up to equal the 100 volt source.

Each voltage drop can be determined by the proportional method by using the size of each load resistance. Load 3 has a resistance value of about 60% of the total resistance. This value works out to 60 volts. Load 1 is 150 ohms, or 15% of the total. This value is 15 volts. Load 2 has a resistance of 250 ohms, or 25% of the total. This value is 25 volts. Either the Ohm's law method or the proportional method can be used.

If one resistance should increase beyond tolerance value in a series circuit, the voltage drop across that resistance would also increase. This would then throw off all of the designed values, and the circuit would not work as it should.

Power in the series circuit is figured in a manner similar to finding other values. Use the power triangle in this case. Take known values from the Ohm's law triangle and transfer these values to their proper places in the power triangle. The result gives the current and the individual voltage drops. Power may be calculated by use of the formula $P = E \times I$. The three power values in the loads of this circuit are 1.5, 2.5, and 6 watts. Total power at the source is 10 watts, or the sum of all individual power values.

Combination parallel-series circuit.   Often in circuit analysis, one finds a circuit that is neither a true parallel circuit nor is it a true series circuit. Looking at a circuit, one often sees that the circuit contains elements that are wired in parallel with each other as well as some elements wired in series. This type of circuit is classified as a combination parallel-series circuit. When attempting to analyze it, one must use the rules that apply for parallel circuits, for the parallel section, and the rules that apply for series circuits for the parts that are in series. There are many different kinds of combination circuits. Some of the simpler ones are shown in Fig. 12-11. In a practical situation, the technician will follow the circuit on the schematic diagram found in the manufacturer's service literature to determine exactly what type of circuit it is.Then he will analyze the voltage, current, and resistance in the circuit, as required, in order to make a correct repair.

The open circuit.   If the definition of an electrical circuit is correct, how would one define an open circuit? The easiest way is to modify the original definition by adding that the open circuit is a circuit that is *incomplete*. Some part of a circuit is missing. It may be either the source, the load, or the lines. Sometimes the open circuit occurs when a control

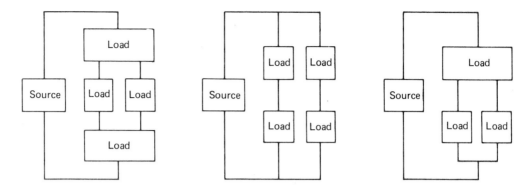

Figure 12-11. Variations of combination series-parallel circuits.

switch is in the *off* position. At other times, the open circuit occurs when any part of the circuit fails or breaks. When this occurs, a very high resistance—almost to infinity—exists at the open point. In electrical work the symbol ∞ is used to indicate that the resistance is *infinite.* The rules regarding resistance and voltage drop across a resistance state that, if the resistance is high, then the voltage drop across that same resistance will also be high. The highest possible value of resistance is an infinite value. Because of this effect, the total applied voltage will appear across an open circuit. Applying this rule, it becomes fairly easy to locate an open circuit in a set. Another way of looking at this same occurrence is to look at the circuit from an electron-flow point of view. If there is an infinite amount of resistance in the circuit, then there will be no electron flow. When there is no electron flow through components in the circuit, there will be no voltage drops developed across these same components. The components act as if they were solid wires as a result of a no-current flow condition. The total applied voltage appears at the open point of the circuit when the open-circuit condition exists.

12-3
Use of Meters
to Measure
Circuit Values

The practical technician uses test equipment much more often than he calculates circuit values. One of the most frequent measurements made is that of circuit voltage drop. Figure 12-12 shows how meters are used to measure common circuit values. *Voltmeter* leads are placed on each side of the area when a voltage drop is to be measured. Since voltage is a difference in electrical pressure between two points in a circuit, the leads to the meter are placed at the two points and the electrical pressure difference is measured. Three voltage measurements are illustrated. The difference is that the point of reference is changed for one measurement. Determine exactly what is to be measured before placing the meter leads in the circuit. Once this is done, then make the measurement. Keep in mind that the voltmeter is placed in parallel with the device across which the voltage drop is measured.

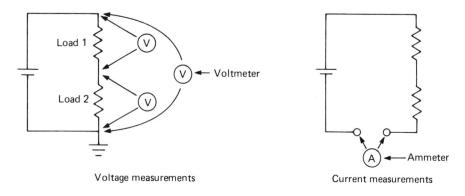

Voltage measurements   Current measurements

Figure 12-12. Placement of meters to measure circuit values.

Current is measured in a different manner. The *ammeter* is connected into the circuit. The circuit is usually broken, or opened, and the meter leads inserted at either side of the open point. Electron flow is through the meter as well as the circuit. The meter, being a flow meter, measures amperage. This amperage is measured at any point of the circuit shown. In a series circuit it should be the same value throughout the circuit.

There is another method of measuring current in a circuit. This method is quite a bit easier than breaking open the circuit and using an ammeter. This method uses a voltmeter and Ohm's law. If the resistance of a component is known and the voltage drop across the component is measured, then these two values are used in Ohm's law to determine circuit current. Some manufacturers are using this procedure on their schematics. Voltage drops across certain components are provided. If the value of voltage drop varies too much from the published value, then there is a problem in that circuit. Further investigation on the components forming the circuit is in order once the abnormal value is found.

SUMMARY

Three factors found in all electronic circuits are voltage, current, and resistance. Voltage exhibits a definite polarity. The voltage may be positive, or it may be negative, *with respect to a common point in the circuit.* A voltage whose polarity constantly changes is called an ac voltage. A voltage whose polarity does not change is called a dc voltage. Voltage forces electron flow. The direction of electron flow is from negative to positive in the circuit. If two or more voltage sources are used, then the values of voltage add up to produce a total voltage. The result may be greater or less than any one of the sources, depending upon the polarity of the sources.

Ohm's law is used to find circuit values. Measurements found for individual components in the circuit are used to calculate the other circuit values, by applying Ohm's law.

A basic electrical circuit consists of a source, a load, and lines connecting the source to the load. There are three basic types of working electrical circuits: the series, the parallel, and the combination series-parallel. Each type of circuit has a set of rules which are applied to it when analysis of the circuit is made.

Voltage drop is a term applied to the value of voltage developed across a resistance when electrons flow through the resistance. A group of two or more resistors connected in series across a source is often called a voltage divider. This is because the applied voltage divides proportionally across each of these resistors.

The proof of the voltage divider rule is found in Kirchhoff's laws. One law states that in a series circuit the sum of the voltage drops (including the source) is zero. Another way of saying this is that the sum of the voltage drops across each load is equal to the applied voltage. Kirchhoff's other law says that the amount of electrons entering a junction are equal to the amount of electrons leaving the junction.

The circuit measurements may be determined by using mathematical equations. They may also be found by using measuring equipment. The most common piece of equipment in use today is the voltmeter. The leads of the voltmeter are connected across the component. The voltmeter measures a difference in electrical pressure between two points in the circuit. The ammeter measures current. It is wired directly into the circuit when current measurements are to be made. An alternate way of measuring current is to measure the voltage drop across a specific component and then to calculate the current in the component. This method eliminates the need to break open the circuit in order to install the ammeter.

QUESTIONS

1. Draw a schematic diagram of a circuit with a common positive point.

2. Two voltage sources are connected in series to the same circuit at the same time. One source is +12 volts. The other source is +3 volts. What is the total applied voltage?

3. What would be the total applied voltage in Ques. 2 if the second source's polarity were reversed?

   Solve the following circuit problems. Show all of your work.

4. Find the total circuit resistance in Fig. Q12-4.

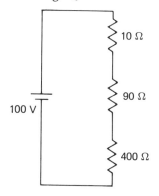

Figure Q12-4.

5. Find the total circuit current in Fig. Q12-4.

6. Find the voltage drops across R_1, R_2, and R_3, in Fig. Q12-4.

7. Find the current through R_1 in Fig. Q12-7.

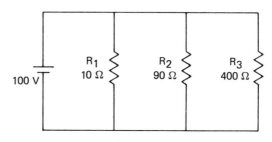

100 V · R₁ 10 Ω · R₂ 90 Ω · R₃ 400 Ω

Figure Q12-7.

8. Find the curent through R_2 in Fig. Q12-7.

9. Find the current through R_3 in Fig. Q12-7.

10. Find the total circuit current in Fig. Q12-7.

Active
Circuit
Devices

13

An active circuit device is one in which an internal operating condition is changed by an outside force while the device is performing work. The opposite is true for passive devices such as resistors, capacitors, and inductors which do not undergo internal change when working in a circuit. The best known active circuit devices are diodes, transistors, and vacuum tubes. These all operate on a similar principle. The principle is that the operating conditions of the device are established by an applied voltage or current. This current or applied voltage is used to control electron movement in the active device. Both the polarity and the amount of the applied voltage or current are critical to the successful operation of the devices.

13-1
Solid-State
Operation

In order to fully understand solid-state device operation, one must spend a little time going over some theory. Most solid-state devices are made from either germanium or silicon. These materials in their purest form are nonconductors. They become semiconductors when a controlled amount of impurity is added to them during the manufacturing process. The type of impurity mixed in with the pure germanium or silicon material determines whether the resulting mixture has excess electrons or a shortage of electrons. Semiconductor materials with an excess of electrons are called N type because they have negative polarity

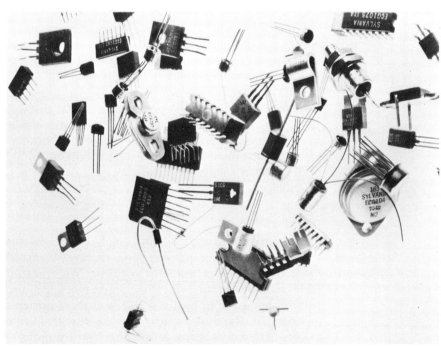

Solid-state and vacuum tube devices used in consumer electronic products. (Courtesy GTE Sylvania, Inc.)

characteristics. Materials with a shortage of electrons are classified as P-type semiconductor material. They accept electrons in a circuit.

This gives us two basic types of semiconductors: the N type and the P type. Both germanium and silicon semiconductors are classified in this manner. Actually, the technician does not normally use either N or P type semiconductors. Practical devices are made up of combinations of these materials into packages called *diodes* or *transistors*.

The solid-state diode. The solid-state diode is a two element device. It consists of an N layer of material and a P layer of material placed together to form a whole, functional unit. Figure 13-1 illustrates how this device works. The diode, in reality, is a voltage-controlled resistance. It exhibits two extreme resistance conditions in a circuit. These conditions depend entirely upon the polarity of the applied voltage. This applied voltage is called *bias*. There are two kinds of bias. These are called *forward* bias and *reverse* bias, referring to the polarity of the bias and to whether it turns the diode on or off electrically. Part (a) of the illustration shows a forward bias condition. The polarity of the bias voltage sets up a low resistance inside the diode. The result is similar to turning on a switch. A lot of electrons flow in the circuit because of the low internal resistance of the diode. Part (b) of the illustration shows a reverse-bias condition. The applied voltage is such that the internal resistance of the diode is very high. This effectively shuts off the diode and very few, if any, electrons flow in the circuit.

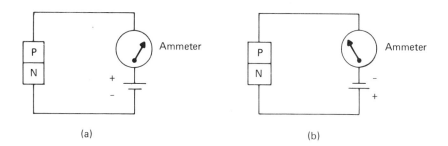

(a)   (b)

Figure 13-1. Conduction in the solid-state diode is dependent upon polarity of the applied voltage.

What we have is a two-condition electron switch. This switch is controlled by the applied bias. There are no moving parts. When the diode is forward biased, it is *on* and electrons move easily through it. When it is reversed biased it is *off* and electrons do not move through it. Each of the elements of the diode has a name. One end is called the *anode*. The other end is called the *cathode*. These are identified by a schematic symbol, as illustrated in Fig. 13-2. Electron flow is in only one direction through the diode. It is from cathode to anode, as indicated by the arrow in the illustration.

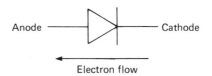

Electron flow

Figure 13-2. Schematic symbol for and electron flow direction in the solid-state diode.

Since both germanium and silicon diodes act as resistors they will show some resistance even when forward biased. A typical germanium diode will have a voltage drop of about 0.3 volt from cathode to anode. The silicon diode has about 0.7 volt from cathode to anode when it is in a forward biased condition in a circuit. When either of these diodes is reverse biased, it has a resistance that is in the megohm range.

When the various factors are put together into a circuit, the diode may easily be used as a switch. Figure 13-3 illustrates this point. A diode is shown connected in series with a 10-kilohm resistance. The resistor and

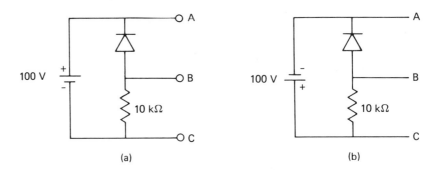

(a) (b)

Bias Condition	Diode Resistance	E Applied (A-C)	E Diode (A-B)	E Resistor (B–C)
Reverse	High	100 V	$\cong$ 100 V	$\cong$ 0 V
Forward	Low	100 V	0.3 V or 0.7 V	99.7 V or 99.3 V

(c)

Figure 13-3. Using the diode as a switch in a circuit.

diode are connected to a 100-volt power source. For the purposes of this illustration, the power source polarity is reversible and it is dc. When a reverse bias condition is established, as in Fig. 13-3(a), the voltage drops developed across the resistor and diode divide, in proportion to the amount of resistance each has. The higher resistance has the larger

196

voltage drop. The drops are proportional to the size of the resistors. In this case the resistance of the diode, being in the megohm range, is at least 1,000 times greater than that of the fixed 10-kilohm resistor. The ratio of voltage drops in this circuit is 1,000 to 1. For all practical purposes all of the applied voltage will appear across the diode and none across the 10-kilohm resistor. The circuit is turned off and there is no electron flow available to perform work.

Reversing the applied voltage polarity establishes a new set of conditions. The diode is forward biased. Its resistance is very low. The voltage drop across the diode is between 0.3 and 0.7 volt. The balance of the applied voltage now appears across the 10-kilohm resistor. Electrons are able to move in this circuit. Work is performed. Assume that the 10-kilohm resistor is actually the load in this situation. The load could be a TV set. When the diode is forward biased, electrons flow through the load and the TV set is working. Reversing the polarity of the applied voltage turns off the TV set.

If the dc source in this circuit is replaced with an ac source, then the set will operate when the polarity of the ac voltage is correct. In other words, the TV set is on half of the time, and off the rest of the time. While this particular circuit is not practical, circuits similar to this are in constant use in electronics. Additional components are required in order to have an operational circuit. These are discussed in detail in the next chapter. The purpose of this presentation is to show how the diode is used as a voltage-controlled switch in a circuit—it actually works as a voltage-controlled resistance, being either a very low resistance or a very high resistance, depending upon the polarity of the applied voltage. When it is on, the major portion of the applied voltage appears across the load and work is performed in the circuit.

The diode is rated, basically, in two ways. One has to do with the amount of current it is capable of handling. The other has to do with the level of voltage it is able to handle when it is reverse biased. The first rating is simply an amperage rating. Specifications for each diode made and classified include the amount of current the diode will allow to move before it self-destructs. Excessive current in any semi-conductor device produces heat. The heat tends to destroy the device. The second rating is called the *peak reverse-voltage,* or PRV, rating. This refers to the amount of voltage that can be applied to the diode in a reverse bias condition. An excess of voltage, when the diode is reverse biased, will break down the insulating qualities of the device and ruin the diode.

Diodes made in the United States use a standardized numbering system. This system uses a sequence of numbers starting with 1N. Typical diodes have numbers such as 1N34 or 1N4002 on them. The 1N indicates a solid-state diode. The other numbers are assigned by the industry. Exact specifications can be found in a manufacturer's reference manual, for the set of numbers used after the 1N indicator. Semiconductors made in countries outside of the United States may use other identification systems. A cross-reference manual provided by a U.S. manufacturer will often help identify operating conditions for these devices.

The zener diode.   One special type of diode is called the zener diode. This diode uses a characteristic found in all diodes. When any diode is reverse biased, very little electron flow occurs. At some point in this situation, the diode starts to break down and to conduct. This point is called the *zener region*. Zener diodes are manufactured to take advantage of this characteristic. The diode is used as a voltage regulator in circuits where close voltage-tolerance is required.

13-2
Transistors

The principles upon which the diode is based are also used for the transistor. The transistor is so named because it operates on the principle of transferring resistance. The transistor is a three-element device. These elements are the *emitter,* the *base,* and the *collector.* There are two kinds of transistors: the PNP (*positive-negative-positive*) and the NPN (*negative-positive-negative*). The transistor is constructed like a sandwich, with the base being the center part. Figure 13-4 shows these two types of transistors and their schematic symbols. The difference between the

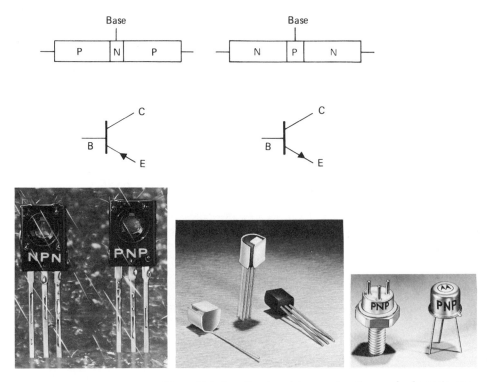

Figure 13-4. Junction-transistor construction, and schematic symbols. (Courtesy Motorola Semiconductor Products, Inc.)

schematic symbols is found in the direction of the arrowhead on the emitter element. The symbol for PNP transistors has the arrow pointing toward the base while that for NPN transistors has the arrowhead pointing away from the base.

The transistor is actually two diodes. The emitter-base diode is normally forward biased in a circuit, while the base-collector diode is reverse biased. Because of the low resistance of the emitter-base diode, it is very easy to turn on the transistor. Transistors are actually controlled resistances in a circuit. There are two circuits involved. One is the input circuit. It consists of the emitter-base diode, the input load, and a bias voltage source. About 5% of the total current moving in a transistor is found in the emitter-base circuit. The other circuit is the base-collector circuit. The collector is biased with a greater positive voltage than the base. The base is a very thin piece of material. As a result, almost 95% of the current moves from the emitter, through the base, to the collector, then to the load, and then back to the power source. This is shown in Fig. 13-5. Variations in base current are used to control current in the collector circuit.

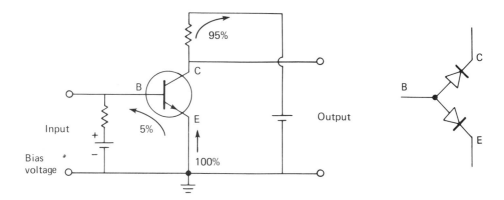

Figure 13-5. Current conduction in the NPN bipolar junction-transistor.

Let's translate this into terms of resistance, as illustrated in Fig. 13-6. A voltage is established at the output terminal of the transistor circuit. The value of the voltage depends upon the internal resistance of the transistor. The internal resistance is based upon the operating condition at the base element. The base voltage controls the amount of current in the emitter-base circuit. A higher voltage increases the current and a lower voltage decreases the current. The emitter-collector circuit acts as two resistances. The load resistance is a fixed value and the transistor resistance varies in relation to the emitter-base voltage. When the base input voltage is low, the transistor has a high resistance. Most of the applied collector voltage appears across the transistor's emitter-collector connec-

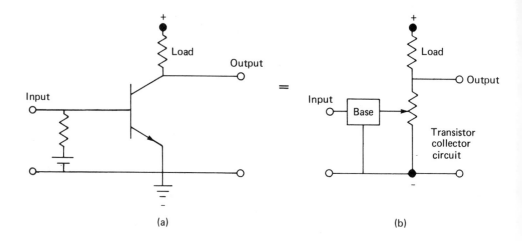

(a)   (b)

Base Voltage	Applied Collector Voltage	E-C Transistor Voltage	Load Voltage	Transistor E-C Resistance
Low	20 V	18 V	2 V	High
Medium	20 V	10 V	10 V	Medium
High	20 V	2 V	18 V	Low

Figure 13-6. Varying the internal resistance of the transistor varies the output voltage.

tion. High resistance in a circuit is associated with low current. Low current values usually mean that little work is done in the load.

Increasing the input voltage and current decreases the internal resistance of the transistor's E-C (emitter-collector) connection. Less voltage develops across the transistor. We know from Kirchhoff's theories that more voltage will appear across the load. A higher load voltage forces an increase in current. The result is that more work is performed. A still higher input voltage causes an even greater reduction in transistor E-C resistance. This forces more current through the circuit and results in more work being performed. It is in this manner that the transistor acts as a controlled variable resistance in the circuit. The emitter-collector connections and the load form a voltage-dividing circuit. The output connection is at the junction of the transistor and the load. The input voltage (or signal) causes a variation in transistor resistance. This results in similar variations in output voltage (or signal) at the load. Applications of this principle, called *amplification,* are discussed in Chap. 14.

200

The transistor circuit discussed in the preceding paragraphs was an NPN transistor. The PNP type works in a similar manner. The major difference is that the polarity of the power source is reversed. Conduction of electrons is in the opposite direction in the PNP. The ratios of 5% and 95% still hold true. The emitter element still handles 100% of the current, as in the NPN transistor. Both input and output circuits function like those of the NPN transistor. Electron flow in the PNP transistor is from the collector, through the base, to the emitter. Electron flow for an NPN transistor is from the emitter, through the base, to the collector. Selection of NPN and PNP transistors is made when the set is designed.

Transistors made in the U.S. have a 2N identifying number. This is followed by a series of two, three, or four digits. Exact operating specifications can be found in a semiconductor manual. Often these transistors are classified with a power rating. Generally, they have either a low, medium, or high power rating. Typical low power transistors are found in tuners and IF sections of radios and TV sets. Where the signal requires further amplification, medium power transistors are employed. The power range for these transistors is from 50 to 100 milliwatts. Output circuits require higher power ratings. Some go as high as 5-50 watts. Power is related to heat. A byproduct of transistor operation is heat. Transistors may be destroyed by too high an operating temperature. Most power transistors are mounted on a metal plate. This plate, called a *heat sink,* is used to conduct the heat away from the transistor and permit a cooler operation. The result is more power in the circuit and less chance of heat damage to the transistor.

The transistor described in the preceding paragraphs is called a *bipolar junction-transistor.* This is because of the construction of the device. Another type of transistor now used in many electronic circuits is called the field-effect transistor.

The field-effect transistor. The field-effect transistor (FET) is of an entirely different type of construction than the bipolar junction-transistor. Conduction in the FET is controlled by means of an electric field rather than by an applied voltage which causes electron flow through the base-emitter junction. This is illustrated in Fig. 13-7. A bar of

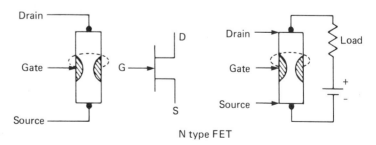

N type FET

Figure 13-7. Field-effect transistor construction, and schematic symbol.

semiconductor material is used as the major conductor of the FET. One end of the material is called the *source*. The other end is called the *drain*. These are connected to a circuit made up of the load and a power source. Electron flow is from the negative terminal of the source, through the FET source connection, to the drain, through the load, and then to the positive terminal of the source. Electron flow is controlled by the *gate*. The gate area is made from opposite polarity material which is diffused into the FET bar during the manufacturing process. The gate area has the ability to enlarge or to contract. This depends upon the amount of voltage applied to the gate during operation of the FET. Increasing the gate area offers a greater resistance area to electrons moving from source to drain. Reducing the gate area increases electron flow through the FET. A constantly varying gate voltage (or signal) will make the electron flow through the FET vary. This, in turn, will change the voltage drop across the FET.

Here, as with the bipolar junction-transistor, the FET acts like a variable resistance in the circuit. The FET and the load make up a series circuit. The two components are working like a voltage-divider circuit. Changing the resistance of the FET changes the voltage drop across it. This in turn affects the voltage drop across the load. The FET is made either as a P-channel or an N-channel device, depending upon the type of semiconductor used. The arrow at the gate indicates the kind of device used in the circuit.

The silicon controlled rectifier.   Another solid-state device used in consumer electronic devices is the silicon controlled rectifier (SCR). See Fig. 13-8. The SCR works in a manner similar to the diode, *except* that it has an additional element. This element, called the *gate,* is used to control the turn-on point of the device. The SCR is made up of four layers of semiconductor material. It is turned on when the anode-cathode operating voltage is reached. It may be turned on at a lower anode-cathode voltage by use of a gate voltage. Once the SCR is turned on, it stays in the on condition until the anode-cathode voltage is removed or the polarity of this applied voltage is reversed. This principle is used in circuits in which

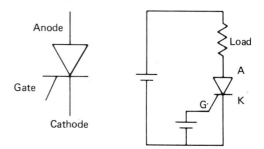

Figure 13-8. Silicon-controlled-rectifier symbol, and a typical circuit.

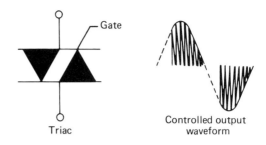

Gate

Triac

Controlled output
waveform

Figure 13-9. Triac schematic symbol, and the resulting output
waveform when the triac is used in a circuit.

a diode with a control element is required. Another common name for
this device is the *thyristor*. The SCR can only be used when the polarity is
correct. This limits its use in an ac circuit to half-time conduction. There
are modified SCRs produced which are made to operate as two SCRs in a
back-to-back arrangement. These are called *triacs*. The symbol for the
triac is shown in Fig. 13-9. The triac is used in ac circuits. It is used to turn
on both halves of the ac signal when the proper gate voltage is applied.

The advantage of the SCR and the triac is that they may be used to
control power in a circuit. A triac output waveform is shown in Fig. 13-9,
also. The amplitude of the waveform is 100% of the value. The on-time of
the wave is reduced to less than 100%. The result is full voltage and
current to the load when on. The total time that power is applied is
reduced. This device finds wide use in light dimmers and motor speed-
control circuits.

Other semiconductor devices. Many other semiconductor devices are
in use in consumer products. They are too numerous to describe in detail
in this book. Mention must be made of some of the more common ones.
The *integrated circuit* (IC) is not actually one device. This is the name used
to generally describe a group of subminiature circuits contained in one of
two or three styles of packages. The IC consists of many transistors and/or
diodes. There are literally hundreds of kinds of circuits currently produced as ICs. Here, as with discrete semiconductors, the user must refer
to a manufacturer's reference manual to find specific information regarding one IC or another. ICs are more and more being used instead of
discrete components in sets.

The *diac* is another semiconductor. It is a two-terminal voltage controlled device. It is used to turn on, or trigger, ac power-control circuits,
particularly those using triacs or SCRs.

The *unijunction transistor* (UJT) is used in trigger circuits, also. It is a
three-terminal device and is used as a control device and in timing
circuits.

There are other types of semiconductors in use and still more on the drawing boards. One of the best ways to stay informed about current developments in electronics is to read any of the many monthly publications available which relate to this field. These magazines and newsletters are very informative and provide initial information very soon after announcement of a new device or discovery.

13-3
Vacuum Tubes

The vacuum tube may be described as a piece of glass tubing from which all air has been removed. It contains two or more metal elements. The elements permit electrons to flow through the tube in one direction. This is because of the materials from which the elements are made. One element, called the *cathode*, gives off electrons, under certain conditions. Another element, called the *plate*, receives the electrons, thus forming a path for electron flow through the tube.

There are two types of cathodes used in electron tube construction: the cold cathode and the hot cathode. The terms relate directly to the condition of the cathode when it emits electrons. Most tubes produced today are of the hot cathode type. Both the cathode and the plate are connected by wires inside the tube to terminals on the base of the tube. These terminals are called *pins*. Hot cathode tubes use a heating element called a *filament* to raise the cathode temperature to the point at which it gives off electrons.

Figure 13-10 illustrates the construction of the two-element vacuum tube and its schematic symbol. The electrical difference between these types of tubes is that one uses a filament as a cathode while the other has a separate cathode. Both perform in the same manner when wired into a circuit. The tube using a separate cathode is classified as an *indirectly* heated tube while the other is classifed as a *directly* heated tube.

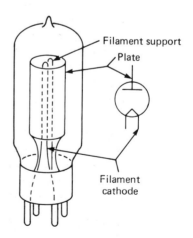

Filament support

Plate

Filament cathode

Figure 13-10. Vacuum-tube diode construction, and schematic symbol.

Electron flow through the tube is from the negative power source, to the cathode, through the vacuum in the tube, to the plate, through the load, and then to the positive terminal of the power source. Conduction occurs only when the plate has a positive charge with respect to the cathode.

Electron control is the two-element tube is achieved by changing either the filament voltage or the plate voltage. This two-element tube is called a *diode*. The filament is not considered to be an active element in the tube. Vacuum tube diodes are rated in a manner similar to the semiconductor diode. Both current handling capability and peak reverse-voltage limits are basic to their ratings.

Tubes are identified in a different way then transistors are. Most tubes are identified by a combination of numbers and letters. For example, a diode tube could be identified as a 5U4GB. In this case, the number before the first letter indicates the nominal operating voltage for the filament. The letter following is not important. The second number indicates the number of elements in the tube. The set of letters at the end is used to indicate the shape of the glass bulb in which the tube elements are housed. Operating filament voltages range from zero to 120 volts. The number of active elements is only limited by industry standards and the physical size of the glass tube. The letters at the end of the designation relate to the specific shape and form of the glass tube. Sometimes the letters A, B, C, etc. are used to indicate a later version of the same tube. A 6S4A tube would be a later version of a 6S4 tube. This later version would probably have some higher specifications than its predecessor. Most tubes used in home-entertainment devices are marked in this manner. Other tubes use only a series of three or four numbers for identification purposes. Tubes using numbers only are normally used in industrial or special purpose applications rather than in consumer products.

The triode tube. Other elements can be added to the diode tube in order to offer some different kinds of control. One element that can be added is called the *control grid*. This grid surrounds the cathode, as illustrated in Fig. 13-11. The tube now has three active elements and is called a *triode*. The purpose of the control grid is to control the electron flow through the tube. This is done by placing a voltage charge on the control grid. If this grid is more negative than the cathode, electrons are not able to pass from the cathode to the plate and the tube is in a state called *cutoff*. Its internal resistance approaches infinity. There is no electron flow.

As the control grid voltage is changed to a less negative condition, electrons are attracted to the plate by the high positive potential on that element. The control grid voltage is varied in order to change electron flow in the tube. This element, the control grid, is actually used to vary electron flow. It is always placed next to the cathode when the tube is designed. Because of the proximity of this grid to the cathode, very little

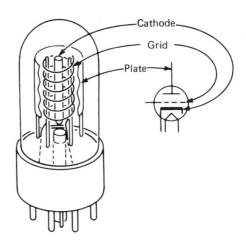

Figure 13-11. Triode tube construction, and schematic symbol.

voltage change is required in order to control electron flow in the tube. When the tube is conducting at its maximum, it is said to be *saturated,* and increasing control grid voltage further will not make any more electrons flow in the circuit.

Multi-element tubes. Vacuum tubes are available with four, five, or more active elements. The four-element tube is called a *tetrode.* This tube uses an additional element. It is called the *screen grid.* The screen grid is located between the control grid and the plate. Its purpose is to help electrons get past the control grid. The tetrode improves the efficiency of the tube.

Another commonly used tube is the five-element tube. This is called the *pentode.* The fifth element, called a *suppressor grid,* is added between the screen grid and the plate. Its purpose is to collect electrons which bounce from the plate during conduction and return them to the plate. This further increases the efficiency of the tube.

Other tube types in use include a group classified as *multipurpose tubes.* These tubes often have two or three electrically separate units housed in one glass container. Some of these tubes are shown in Fig. 13-12. The available combinations seem to be limited only by the imagination of those designing the tubes. Exact electrical specifications, typical operating conditions, and applications can be found in a tube manual. The manual provides base pin connections, physical size, and all other pertinent information. Each tube manufacturer publishes a tube manual for use by technicians.

One other tube type should be mentioned at this time. There is a group of tubes called *gas tubes.* They contain a gas which ionizes at a predetermined voltage. The ionized gas conducts electrons at a constant rate. These tubes are used as rectifiers and voltage regulators in a manner

206

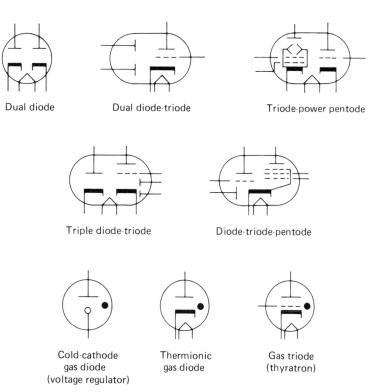

Dual diode   Dual diode-triode   Triode-power pentode

Triple diode-triode   Diode-triode-pentode

Cold-cathode   Thermionic   Gas triode
gas diode   gas diode   (thyratron)
(voltage regulator)

Figure 13-12. Schematic symbols for multipurpose vacuum tubes and gas filled tubes.

similar to the SCR and the zener diode. They are identified on the schematic by a black dot in the lower right corner of the schematic symbol. Their use in consumer electronic devices is very limited at the present time.

13-4
The Active Device as a Controlled Resistance

Both tubes and semiconductors act as controlled variable resistances in any circuit. This controlled resistance could be achieved by using a variable resistor. However, in most cases, the time taken to perform the work would be too great. The advantages of using the active device are speed and accuracy. Figure 13-13 illustrates these devices. All of the devices shown are used as variable voltage-dividers in their respective circuits. Keep in mind that circuits other than those illustrated here are used in electronics, too. Each of the devices shown uses some method of adjusting the amount of resistance in the device. The variable resistor uses a mechanical means. The junction transistor uses an electron current

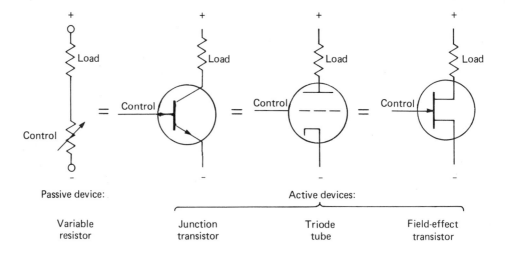

Passive device: Active devices:

| Variable resistor | Junction transistor | Triode tube | Field-effect transistor |

Figure 13-13. Comparision of passive and active circuit devices.

signal. Both the triode tube and the FET use a signal voltage. Regardless of the method employed, each control system sets up the amount of internal resistance for the active device.

The power source is connected to the load resistance and the negative terminal of the device. This establishes a voltage-divider network consisting of the load resistance and the device. Voltage drops are developed across the load and the device in proportion to their resistance. When the device has a high resistance, the voltage drop across it will also be high. When the device has a low resistance, the voltage drop across it will also be low. Varying the resistance at the control causes the voltage across the device to vary, also.

In most circuits the amplitude of the control voltage or current is much less than the variation in voltage developed across the controlled device. In many of these circuits the shape of the input signal is reproduced across the device. This produces a magnified waveform. Another term for this is *amplification.* Transistors and tubes act as amplifiers in most applications. These applications are fully discussed in the next chapter.

SUMMARY   The active circuit device is one whose internal resistance is changed by an outside force. Active devices are generally either solid-state semiconductors or vacuum tubes. Common semiconductors are the diode and the transistor. Typical vacuum tubes are the diode, triode, and pentode. Each of these is used as a controlled resistance in a circuit, so that some electrical work is performed.

The semiconductor diode consists of two kinds of material. These are P-type and N-type materials. When placed together they form a diode with a junction. The desirable characteristic of the P-N junction is that it offers a low resistance to the circuit under certain applied voltage conditions, and it offers a high resistance to the circuit under other applied voltage conditions. This is called forward bias when the resistance is low and reverse bias when the resistance is high.

The diode acts as a two-condition switch under these conditions. It is either *on* or *off*, depending upon the polarity of the applied voltage. Diodes are rated both by current handling capability and by the value of the voltage handled when reverse biased. Most diodes made in the United States are identified by a number sequence starting with 1N.

One special diode is the zener diode. This diode is used as a voltage regulation diode in circuits. It is normally connected in the circuit in a way that is the reverse of other diode wiring.

The transistor is a three-element semiconductor. Commonly used transistors include the bipolar junction-transistor and the field-effect transistor. Both have an element which is connected to the power supply return. They also have another element which is connected to an outside voltage or current source. This element, called the base or gate, is used to control conduction through the transistor.

Bipolar junction-transistors are produced as either NPN or PNP types. This designation refers to the construction of the transistor. The emitter and collector are made of material of the same polarity while the base is made of opposite polarity material. These transistors operate with 100% of the conduction in the emitter—about 5% in the base-emitter circuit and about 95% in the emitter-collector circuit. This holds true for both NPN- and PNP-type transistors.

Field-effect transistors are of either P-channel or N-channel construction. This designation refers to the polarity of the source-drain material. The gate element is made from opposite polarity semiconductor material. Most transistors produced in the U.S. have a 2N identifying number followed by two to four digits.

A characteristic of the semiconductor is the production of heat as conduction occurs. Too much heat will destroy any semiconductor. Diodes and transistors have to be operated at a temperature below that at which they will be destroyed. One method of cooling the device is to mount it on a metal surface called a heat sink. Heat from the device is conducted to the metal plate and then this larger surface is cooled by the surrounding air.

Another semiconductor device finding increasing application in consumer electronic products is the silicon controlled rectifier. The SCR acts like a diode except it has a trigger element in it. This permits the SCR to be turned on for only a part of its conduction cycle. The result is full power for shorter time periods. SCRs are used in timing circuits for TV, light dimmers, and motor speed-controls.

Another family of active devices is the vacuum tube family. The vacuum tube operates on the principle that electrons are emitted from a metallic surface when that surface is heated. Another surface in the tube is charged positively and collects the electrons. These elements are called the cathode and the plate of the vacuum tube. The cathode is heated by a filament in the tube. Conduction occurs in only one direction through the tube. The tube acts in a manner similar to that of the semiconductor diode.

Three-element tubes are called triodes. The third active element is called the control grid. It is placed between the cathode and the plate of the tube. Its purpose is to control the electron flow through the tube. This is accomplished by means of a voltage applied between control grid and cathode. This control-grid voltage is so effective that it is able to stop all electron flow through the tube.

Additional elements that may be added to the tube are the screen grid and the suppressor grid. Each is used in order to make the tube a more efficient conductor. A four-element tube is called a tetrode, and a five-element tube is called a pentode. In many cases two or more sets of elements are built into the same tube. These multipurpose tubes are very common today. Exact specifications and operating conditions for both tubes and semiconductors are to be found in manuals published by the manufacturers of these devices. Some manufacturers provide cross-reference material to assist in selection of the proper component.

QUESTIONS

1. What is meant by the term *forward bias*?

2. Explain forward and reverse bias in relation to internal resistance of the diode.

3. Draw the schematic symbol for a semiconductor diode. Show direction of electron flow.

4. Show how a diode may be used as a switch.

5. Explain how the transistor works as a variable resistance.

6. What is the difference between an NPN and a PNP transistor?

7. How does the output circuit of a transistor or tube act as a voltage-divider network?

8. What is the difference in operation between an SCR and a triac?

9. Name the elements of a pentode tube and explain their function.

10. How does the active device work as a controlled resistance in the circuit?

Active
Device
Applications

14

Twelve items basic to the understanding of electronic theory are identified earlier in this book. These include the three components called the resistor, the capacitor, and the inductor. They also include the three concepts of resistance, voltage, and current. Series, parallel, and combination series-parallel circuits are also identified. These nine items are presented in earlier chapters. Three items remain to be discussed. These three are related to circuits which use active devices. Once these three circuits are understood the technician only has to learn about some of the specialized circuits. The basic circuits using active devices are the *rectifier*, the *amplifier*, and the *oscillator*. The actual circuitry is discussed in this chapter in order to help the reader understand how the circuits work and where they are used in consumer electronic products.

An introductory book of this kind cannot possibly delve into all of the circuits used in electronic devices. The intent here is to present basic ideas and show how these are used in some of the more common applications. Most of the material presented shows how the blocks used in electronics are assembled and how the various signals are processed through these blocks. Many excellent books are available which go deeply into the topic of circuit analysis. One of the purposes of this book is to whet the reader's appetite for more knowledge relative to the subject of electronics. This chapter tells how basic circuits with active devices work. This fundamental knowledge may be used to start one on the way to a full, in-depth, understanding of electronic circuitry.

Electronic devices using transistors and tubes require a dc power source. Most radios, TV sets, and stereos have ac power cords and are connected to the ac power outlets in the home. In order for these sets to work, this ac is changed into dc voltages in the set's power supply section. Conversion of ac to dc is done in several steps. These steps are related to blocks in the set. Figure 14-1 illustrates these blocks. The ac is obtained from the wall outlet. The necessary voltages are then obtained from a transformer. The ac voltage at the output of the transformer is then *rectified,* or changed into pulses of dc. In order to obtain pure dc for use within the set, the pulses are processed through a filter. The purpose of the filter is to smooth out any variation in the dc voltage. The output of the filer is a pure dc voltage. In some power supplies an additional block is used. This block is called a *voltage regulator.* Its purpose is to keep the dc voltage at a fixed value. This is done to compensate for any changes which might occur in input voltage or for variations in load circuit resistance.

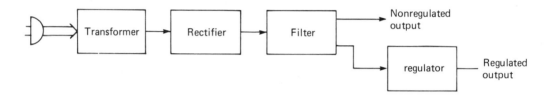

Figure 14-1. Block diagram of a power supply with regulated and nonregulated output.

The halfwave rectifier. There are several kinds of rectifier circuits used in electronic devices. The simplest of these is called the halfwave rectifier, so called because the circuit rectifies only half of the ac input waveform. This is illustrated in Fig. 14-2. The ac input voltage may be supplied

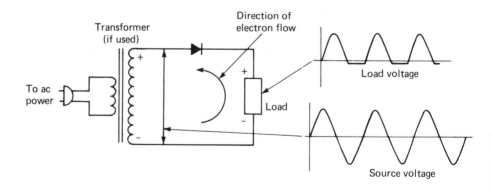

Figure 14-2. The halfwave rectifier circuit, with input and output waveforms.

directly from the power line. It may be sent through a transformer in order to step it up or down to the proper value for use in the set. Note the difference in waveforms at the source and at the load. Follow electron flow in this circuit. Electrons start from the negative point and flow through the load, the diode, and back to the transformer source. The polarity of the voltage at the transformer is such that the diode is forward biased. When the diode is forward biased, most of the applied voltage develops across the load. Earlier discussions showed that a forward biased diode develops either 0.3 or 0.7 volt across its junction. The balance of the applied voltage develops across the load.

When the polarity of the ac voltage reverses, as it does periodically, the diode is reverse biased. At this time, the diode resistance is very high and most of the applied voltage appears across it instead of across the load. The resulting output voltage looks like half of the input waveform, thus the name *halfwave rectifier* for this circuit.

Pulsating dc is not too useful in a circuit. Most circuits require a pure dc voltage. The filter network helps smooth out and eliminate any variations in the dc voltage. The level of dc voltage is above that of the zero-voltage reference line in Fig. 14-3. There is some amount of variation in this voltage. It is called *ripple voltage*. This term refers to the small variations in the level of dc voltage at the output of the filter circuit. The amount of ripple voltage is dependent upon the action of the capacitors in the filter network. A small amount of ripple, usually less than 5% of the total dc voltage, is acceptable for most electronic circuits.

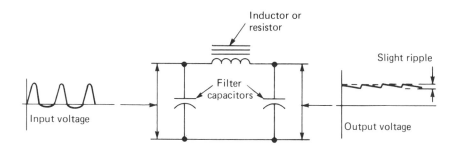

Figure 14-3. Filter action. The network, consisting of the capacitors and inductor, smooth out the pulsating dc.

Vacuum tube circuits operate in a manner very much the same as semiconductor circuits. The diode tube replaces the solid-state diode. The tube circuit also requires a power source to heat the filament of the tube. The circuit for a halfwave rectifier using a tube is shown in Fig. 14-4. When the plate of the tube is positive with respect to the cathode, conduction occurs—as when a semiconductor is used. The output waveform is also a halfwave in this circuit.

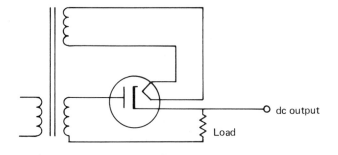

Figure 14-4. A vacuum tube halfwave rectifier-circuit.

The fullwave rectifier. The fullwave rectifier uses two diodes in the circuit. Both a solid-state and a tube circuit are shown in Fig. 14-5. Follow the electron flow through this circuit. Start at a point which will have the top of the secondary windings of the transformer negative. The center tap, or connection, of the transformer is the common point of this circuit. The transformer when connected in this manner acts like it has two

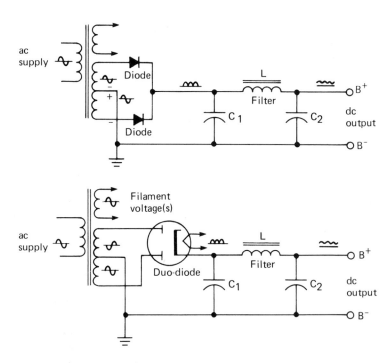

Figure 14-5. Fullwave rectifier circuits.

separate secondary windings. Each has its own polarity. Starting at the top, it would be positive. The lower end of the upper winding (connected to common) is negative. The upper end of the lower winding is positive and the lower end of the lower winding is negative. (The polarity symbols are shown on the schematic.) What this does for the circuit is to forward bias one diode and reverse bias the other diode. Electron flow is from common, through the load, through the upper diode and back to the transformer. Since the other diode is reverse biased at this time, it does not conduct. The output voltage is a halfwave. Now, reverse the polarity of the applied voltage. With an ac voltage this occurs often, depending upon the frequency of the applied voltage. If the source is connected to the ac house line, then the frequency is 60 hertz. Some currently produced TV sets use the horizontal frequency of 15,750 hertz as a secondary ac source. The principle is the same and the results are similar.

When the source voltage is reversed, the polarities at the transformer secondary are also reversed. Positive changes to negative, and negative to positive, at each point. This changes the bias on the two diodes. The lower diode is now forward biased and the upper diode reverse biased. The electron flow is now from the common, through the load, through the lower diode and back to the transformer. The output voltage is in halfwave form. There has been a difference in time for conduction in the diodes in this circuit. They are staggered in relation to each other. All electron flow is through the load, however. The output waveform is shown in Fig. 14-5, also. There are twice as many pulses of dc in this circuit as there are in a halfwave circuit. The pulsating dc is processed by the filter circuit. The result is a dc voltage with little or no ripple on it. This type of circuit is found in many consumer electronic products. It requires a transformer with a center tap connection in order to work.

The fullwave bridge circuit. Another rectifier circuit is the fullwave bridge rectifier. This circuit is shown in Fig. 14-6. The transformer used in this circuit does not have a center tap connection. Electron flow occurs in the same way as in other circuits. The diodes help *steer* the electrons through the circuit. Start when the bottom of the transformer's secondary winding is negative. Electron flow is from the transformer, through diode "d," through the load to diode "a," and back to the transformer. Only two diodes are used. Reverse the polarity of the applied voltage. Electron flow is now from the top of the transformer's secondary winding, through diode "b," through to the load, through diode "c," and back to the transformer. Electron flow is in one direction through the load. The output waveform is fullwave pulsating dc. It looks just like the waveform from the fullwave rectifier circuit using two diodes. The pulsating dc is then filtered.

The voltage doubler. There are times when higher voltages than are available from the source are required. One way of obtaining these voltages without the use of a step-up transformer is by use of a voltage multiplying circuit. Some of these circuits are able to increase the applied

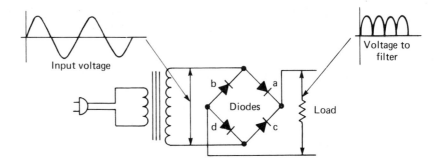

Figure 14-6. The fullwave bridge rectifier-circuit.

voltage by two, three, four, or five times the input voltage level. They work on the principle that a capacitor can be charged and then this charge used to supply additional voltage to the circuit, almost like an additional source. Figure 14-7 shows a halfwave voltage-doubler circuit. Follow the electron flow through this circuit for one complete cycle of the input ac wave.

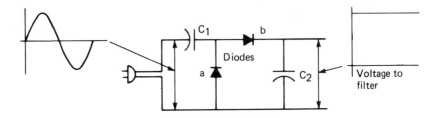

Figure 14-7. A voltage-doubler circuit.

Start when the upper connection to the ac source is negative. Electron flow charges C_1 and moves through diode "a" back to the input. C_1 is charged with nearly the source voltage. Next, the polarity of the source voltage is reversed. Electron flow is from the source, through the load, through diode "b," and back to C_1 which now acts as a part of the source. The voltage stored in C_1 adds to the source voltage. This voltage charges C_2, which is in parallel with the load, to twice the source voltage. The output voltage is fed to a filter network where it is smoothed and used to operate the set.

Diode detectors. Another common use for the rectifier circuit is the detection of signal and the removal of the signal from its carrier. The simplest form of this is the halfwave detector circuit used in the AM radio. This circuit is illustrated in Fig. 14-8. TV video circuits, being amplitude modulated, also use this circuit. The diode and the load's resistor form a series circuit. Electron flow is from the IF amplifier, through the load,

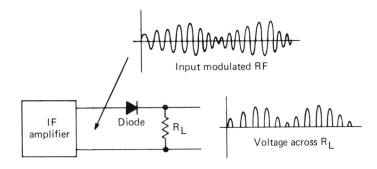

Figure 14-8. A diode detector circuit. The resulting wave is filtered, leaving only the audio wave shape.

through the diode, and back to the IF amplifier. Reversing the elctron flow literally cuts off any voltage developed across the load. The result of this action is a halfwave signal made up of the half of the IF carrier with the amplitude modulation form. The carrier is filtered out, leaving the lower frequency variations which form the signal. The signal is then fed to the next stage for processing.

The FM detector circuits used in FM radios and for the audio portions of the TV signal use a different kind of detection system. This system is frequency sensitive rather than amplitude sensitive. One kind of detector is called a *discriminator*. It works on the principle that changes in frequency will produce a small voltage at the output of a tuned transformer. This is illustrated in Fig. 14-9. Note that the secondary is tuned to three frequencies: 10.6, 10.7, and 10.8 megahertz. There are two diodes in this circuit. The center tap of the transformer is not connected to common.

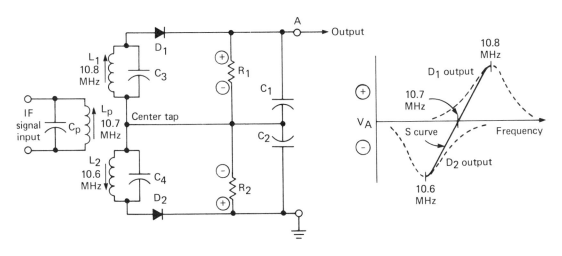

Figure 14-9. Discriminator action. Frequency variations produce slight voltage variation across the diodes.

Only a signal at the IF frequency of 10.7 MHz sets up a balance across both diodes and load resistors. The result is an output voltage of zero. When the audio causes the IF frequency to shift upward, a positive voltage develops at the output terminals. A downward frequency shift produces a negative voltage at the output. These in turn produce an audio signal voltage of varying amplitude and frequency which duplicates the original modulating signal.

Filter circuits. The purpose of the power supply filter is to smooth out the waveform by removing the ripple. Both passive and active filters are used in consumer electronic devices. *Passive* filters consist of capacitors and inductors or resistors. The capacitor is connected in parallel with the rectifier output and load. This is shown in Fig. 14-10. A *pi* section filter consists of two capacitors and an inductor. It resembles the Greek letter π. The second circuit is called a choke input circuit because the choke is placed in the circuit after the rectifier. A capacitor input circuit has the capacitor before the inductor and does not use the second capacitor. Action in the circuit depends upon the charging and discharging of the filter capacitors. When the voltage rises the capacitor charges. This tends to hold the voltage down. When the voltage drops, the capacitor discharges into the circuit and tends to keep the voltage up. The amount of charge depends upon the capabilities of the capacitor.

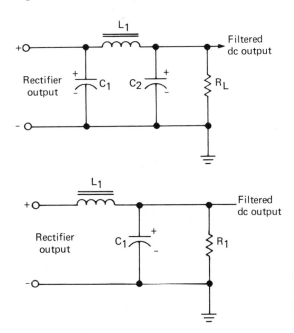

Figure 14-10. Two common types of inductance-capacitance (LC) filter networks. (The upper figure is of a *pi* network. The lower figure is of a *choke input* network.)

The other type of filter used in electronic devices is an *active* filter. It utilizes a transistor to aid in the regulating process. The reason for using a transistor instead of capacitors is that the transistor reacts to compensate for variations in voltage much more rapidly than does a capacitor. The low values of dc voltages used in solid-state sets require large amounts of capacitance in order to filter out the ripple. The active filter does the same work with smaller parts and probably at less cost than the larger capacitors. It is shown in Fig. 14-11.

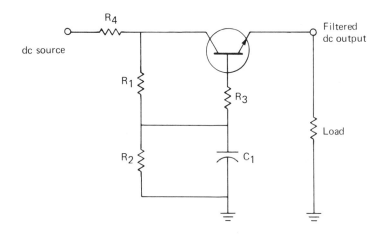

Figure 14-11. The active power-filter circuit.

In that circuit electron flow is from common through R_2, R_1, R_4 and then to the dc source. The three resistors make up a voltage-divider network. This network sets up a bias at the base of the transistor. C_1 is used to maintain a steady dc voltage at R_3 when it is in parallel with R_2. A second circuit is established from common, through the load, then through the emitter-collector of the transistor and R_4 to the dc source. Ripple at the dc source causes small changes in the emitter-base circuit. These change the emitter-collector resistance in order to remove the ripple.

Voltage regulation. Two kinds of voltage regulator circuits are found in entertainment devices. One is the zener regulator and the other is an electronic regulator. The zener diode action is explained in an earlier chapter. The circuit for the zener regulator is shown in Fig. 14-12. Unregulated dc is connected to a series circuit consisting of a regulating resistor and the zener diode. The load is connected in parallel with the diode. The zener diode is designed to regulate at 12 volts in this circuit. A voltage higher than 12 volts is applied to the circuit. Electron flow is from source negative, through the zener, through the regulating resistor, and back to the source. The zener conducts at 12 volts, maintaining this voltage. The

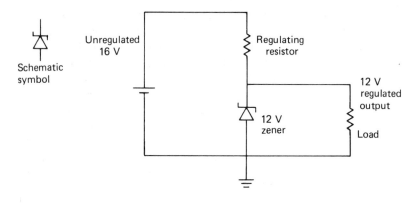

Figure 14-12. A zener-diode voltage-regulator circuit.

balance of the applied voltage develops across the rest of the voltage divider which is the regulating resistor. The output voltage at the load is held to a constant 12 volts by this circuit.

There are several types of electronic regulator circuits in use today. One of these is the series regulator. It is shown in Fig. 14-13. Two transistors are used in this circuit. One is the regulator transistor and the

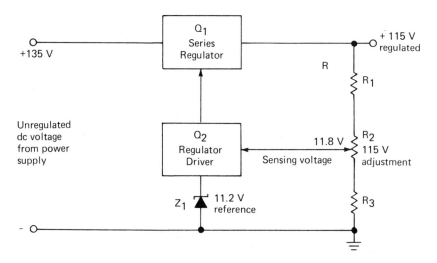

Figure 14-13. A voltage regulator circuit, in its simplest form.

other is used as a control transistor. Many of the actual circuit components have been removed in order to highlight the principle upon which this circuit operates. Electron flow is from source negative, through the voltage divider network consisting of R_3, R_2, R_1 and the emitter-collector of the regulator transistor, and then back to source positive. The control transistor has two reference voltages. One is established by the zener

220

diode at the emitter of the transistor. The other is established by R_2 and feeds the base of this transistor. The collector is connected to the base of the regulator transistor. Differences in these two voltages from those established when the circuit was adjusted will cause the regulator transistor conduction to change. This changes the various voltage drops in the voltage-divider network and maintains the desired dc output voltage.

It is possible to use tubes as voltage regulators, in voltage multiplier circuits, and as bridge rectifiers. Before the use of the semiconductor, tubes were more commonly used as rectifiers. Very few consumer electronic devices use the tube in low voltage power-supply circuits. Some tubes are used as high voltage rectifiers in TV sets, and tubes are used as regulators in the high voltage circuits of color TV sets. Electron flow is still the same through the tube regardless of the circuit application. Use the principles explained in this section to see how all of these circuits work.

14-2
Amplifiers

The amplifier, by definition, is a device which uses a small amount of power to control a larger amount of power. Both transistors and vacuum tube amplifiers are used in consumer electronic products. The video signal as received by the antenna of the TV set is almost 0.00005 volt. The picture tube requires about 50 volts of signal. The signal is increased in strength about 10 million times in the set in order for us to see a quality picture. This is done by amplifiers in the set. Transistor and tube amplifiers both use the same concept of operation. Each has an input circuit and an output circuit. These are illustrated in Fig. 14-14. Each uses a small

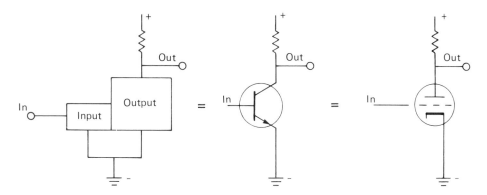

Figure 14-14. Amplifier circuits. Block diagram, with related transistor and tube circuits.

amount of power to control a larger amount of power. One of the three "legs" is common to both input and to output. Look at Fig. 14-15. This illustrates a transistor amplifier with the emitter common to both the input and the output portions of the circuit. Resistors R_1 and R_2 are used as a voltage divider to establish a fixed operating bias point for the circuit.

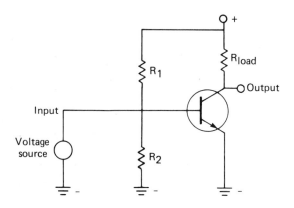

Figure 14-15. The common-emitter amplifier-circuits.

There are two paths for current in this circuit. One is from common, through the emitter, to the base, to the input signal voltage source. This is called the input circuit. The input signal voltage adds to the fixed bias at the base of the transistor. The result is a slight change in base voltage. Base voltage controls base current. In effect, the changes in base voltage control current in the emitter-base circuit. This is the input portion of the circuit.

The output circuit conduction is from common, through the emitter, past the base, through the collector, through the load, to the positive source terminal. Since 95% of conduction occurs in this path, this becomes the larger power portion of the circuit. The transistor, as previously stated, acts like a variable resistance in the circuit. The transistor and the load form a voltage dividing network. The voltage drop across the transistor will vary as its resistance changes. When the transistor has a high resistance it will also have a high voltage across it. When this is true, the load will have a low voltage across it at the same time.

The input signal is normally a very small value. This is used to control much larger values in the output circuit. In this way, the small amount of power in the input section controls a larger amount of power in the output section, and the transistor acts as an amplifier. Technicians refer to these circuits as amplifiers, but they are actually controlled variable voltage-divider circuits. Figure 14-16 illustrates how this works. The values given are not necessarily working values. They are offered as illustrations of how the device works. Actual values are available in transistor manuals showing operating conditions for specific devices. The chart shows output operating conditions for various levels of input signal. Dc is used as the input signal. This signal ranges from 5 to 15 microamps for an overall change of 10 microamps. The 10 microamp change is produced by a change in voltage of 1 volt. In the output circuit, the emitter-collector current ranges from zero to 4 milliamps. This produces a voltage drop across the transistor which is the reverse of the applied voltage and

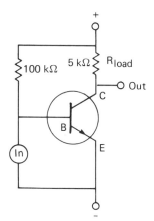

Input Conditions			Output Conditions			
E-B Volts	E-B Current	E-B Power	E-C Volts	Load Volts	E-C Current	E-C Power
1.5	15 µA	22.5 µW	0	20	4.0 mA	0.0 mW
1.4	14 µA	19.6 µW	2	18	3.6 mA	7.2 mW
1.3	13 µA	16.9 µW	4	16	3.3 mA	13.2 mW
1.2	12 µA	14.4 µW	6	14	2.8 mA	16.8 mW
1.1	11 µA	12.1 µW	8	12	2.4 mA	19.2 mW
1.0	10 µA	10.0 µW	10	10	2.2 mA	22.0 mW
0.9	9 µA	8.1 µW	12	8	1.8 mA	21.6 mW
0.8	8 µA	6.4 µW	14	6	1.4 mA	19.6 mW
0.7	7 µA	4.9 µW	16	4	1.0 mA	16.0 mW
0.6	6 µA	3.6 µW	18	2	0.5 mA	9.0 mW
0.5	5 µA	2.5 µW	20	0	0.0 mA	0.0 mW

Figure 14-16. Operating conditions for the common-emitter amplifier circuit.

current. It ranges from 20 to zero volts as the input voltage changes. What is happening is that increases in input voltage cause the transistor's internal resistance to decrease. Compare the input voltages on the charts with their respective output voltages. Operation of the transistor becomes less of a mystery if you keep this reverse resistance idea in mind.

An important characteristic to look at in the chart is the relationship of the power columns. Compare the differences in power for the first two lines. There is a change of 2.9 microwatts of power. These 2.9 microwatts control a charge of 7.2 milliwatts of power in the output circuit. This gives proof for the statement that a small amount of power is controlling a larger amount of power in this amplifier circuit.

The discussion so far has been concerned with a dc input and the relationship of this input to the dc voltage between the output terminal of the circuit and common. This same principle works with an ac signal or voltage as the control. Look at the chart and assume that the input signal voltage is varying at a constant rate. Assume that it looks like a sine wave. If this is the case then the output voltage, which is dc, is changing as the input changes. The output voltage at the collector of the transistor takes on the sine wave shape. Even though it is still dc, it looks like and acts like ac in the circuit. In most amplifier circuits, amplification occurs between the two extreme operating conditions called *cutoff* and *saturation*. The dc at the output will, in most cases, look like the input signal except that it will be larger in size.

This holds true for tube and FET devices as well as for the junction transistor. The basic difference between the junction transistor and the other two devices is that the tube and FET use a *field effect* to control current through these devices. Instead of having an input circuit with electron flow, an applied voltage establishes an *electrostatic field* in the

223

device. An application of this is shown in Fig. 14-17. There is no path for current in the input circuit. The input signal voltage sets up an electrostatic field at the control grid or at the gate. The size of this field determines the amount of internal resistance of the device. The output circuit is a voltage-divider network made up of the device and the load. It operates just as the junction transistor does.

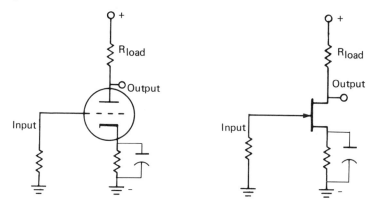

Figure 14-17. Field-effect transistor and vacuum-tube triode amplifier-circuits.

Amplifier classification. Amplifiers are classified in at least five different ways. They are classified according to how the amplifier is used, how it is biased, how signals are coupled from one stage to another, the circuit configuration, and the operational frequency range. Many amplifier circuits fit into more than one classification. The name given for the manner in which the amplifier works in the circuit helps to identify the proper group of classifications.

Use. When amplifiers are classified by their use, they fit into two general groups: voltage amplifiers and power amplifiers. The voltage amplifier is used to increase the amplitude of the signal voltage. This amplifier acts as a voltage dividing network made up of the transistor or tube amplifier and the load resistance. The amount of signal voltage developed across the load compared with the input signal amplitude determines the amount of amplification in each stage. Power amplifiers, on the other hand, are designed to produce a large amount of current in their output load. Power amplifiers are also known as current amplifiers. They may be considered as current control stages.

Another type of amplifier is the *buffer* amplifier. Its purpose is to isolate one stage from another stage. Interaction between two stages is minimal when a buffer amplifier is used between them.

Bias. An amplifier is also classified according to the bias on it. This means that it is classified according to the portion of the input signal voltage cycle during which conduction occurs. Grouped this way, there are four classes of amplifiers: class A, class B, class AB, and class C.

Class A amplifiers are biased so that they are always on. Conduction occurs during the entire portion of the input signal. The result is a minimum of distortion of the output wave. Class A amplifiers faithfully reproduce the shape of the input signal.

Class B amplifiers are biased at the cutoff point. The resulting output waveform is very similar to that of a halfwave rectifier without any filter. When the device is on, it may operate at higher than normal power ratings. When it is not conducting it has an opportunity to cool. Since it is on only half of the time it can cool the other half of the time. When it is on, it may run at higher than normal current values in order to produce higher power ratings.

Class AB amplifiers are biased to be on more than half of the input cycle, but for less than the complete cycle. These amplifiers are used as a compromise between the higher power available from class B and the lack of distortion from a class A amplifier.

Class C amplifiers are biased beyond the point of cutoff. Output conduction occurs only during the positive peaking of the input signal cycle. These amplifiers have high power outputs. They are primarily in RF output circuits.

Coupling. There are three kinds of circuits used to transfer a signal from one stage to another. These are the *resistance-capacitance* (RC), *transformer*, and *direct-coupled* circuits. They are illustrated in Fig. 14-18. The name refers to the component used to pass the signal on to the next stage. The RC coupling circuit uses voltage variations on one plate of a capacitor to transfer changes in signal values to the other plate of the capacitor. The dc level of the collector voltage is not coupled to the base of the next stage in this arrangement.

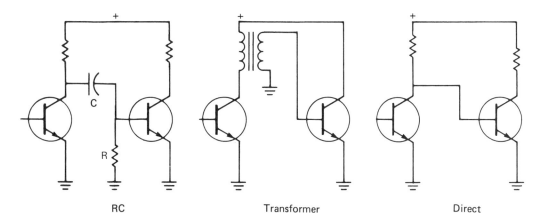

RC   Transformer   Direct

Figure 14-18. Methods of transferring signals from one circuit to another are resistance-capacitance, transformer, and direct coupling.

Another method of coupling a signal is by use of a transformer. Changes in current in the primary of the transformer induce a voltage in the secondary. There is no direct wiring between windings of the transformer. The induced voltage has the shape of the signal.

The third method is called direct coupling. Here, a wire is connected directly from the output of one stage to the input of the next stage. Changes in dc operating voltages at the emitter are used as the operating bias at the base of the following stage.

Circuit configuration. Still another way to classify amplifiers is to identify the element of the active device which is connected to the signal common in the circuit. An amplifier can be considered as a device having two input terminals and two output terminals. One input and one output terminal are wired to a common point. Considered this way, amplifiers are classified as common emitter, common collector, and common base amplifiers. The complete range of combinations for classifying amplifiers in this manner are shown in Fig. 14-19. In the illustration, each circuit is shown, as well as the input and output connections. Input and output signals are also illustrated.

For example, the type of amplifier circuit used more often than any other is the group called common emitter, common source, or common cathode. The *common emitter-source-cathode* circuit provides the largest voltage gain of the three types of circuits. The input to this circuit is connected between the base, gate, or control grid and the common element. The output is connected between the collector, drain or plate and the common element. This circuit inverts the input signal. The output signal is 180 degrees out of phase with the input signal. This is the most commonly used amplifier circuit.

Another circuit is the *common collector*, drain, or plate. The input in this circuit is between the control element and the common element. The output is connected between the emitter, source, or cathode and the common element. This circuit is used in output stages. It is very useful when a low voltage, high current signal is required. This circuit does not invert the output signal. It is the same phase as the input signal.

The third circuit is the *common base*, gate, or control grid circuit. The input to this circuit is between the emitter, source, or cathode and the common. The output is connected between the collector, drain or plate and the common element. This is the least used of the three amplifier circuits. The output signal is also in the same phase as the input signal in this circuit.

Frequency. One of the most common ways of labeling amplifiers is according to the specific group of frequencies being amplified. Some of the more familiar terms used to describe these circuits include audio frequency (AF), intermediate frequency (IF), radio frequency (RF), and video frequency amplifiers.

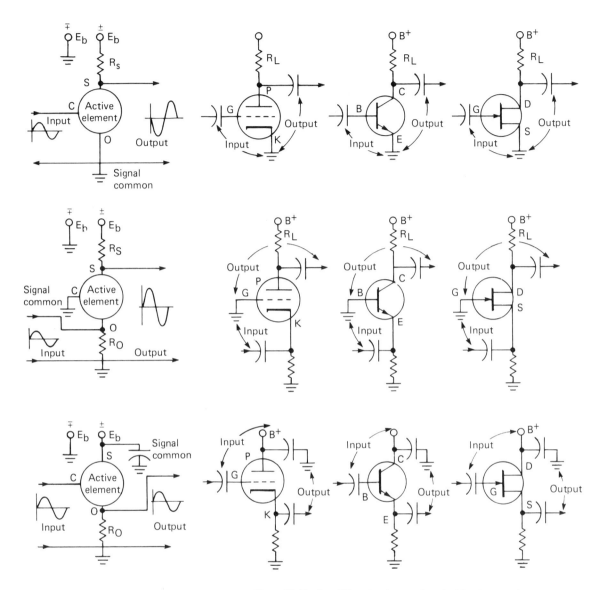

Figure 14-19. Amplifier circuits are classified by the signal-common element. These include common emitter, common base, and common collector for bipolar transistor circuits.

These terms may be used along with the others discussed in this section when identifying a particular amplifier circuit. For example, the expression "Class A, voltage, RC coupled, RF amplifier" could be used to correctly describe a specific circuit. Usually, this would be shortened to become "RF amplifier." The rest of the description would most likely be dropped as too cumbersome.

An oscillator is a signal-generating circuit. It is possible to use a mechanical generator to produce an ac signal. This device is limited to lower frequencies. It would self-destruct at high rates of speed due to centrifugal forces acting upon it. The electronic oscillator is limited only by the components used to construct the circuit. Basic requirements for an oscillator circuit include an amplifier circuit, a frequency determining circuit, and a feedback path. Oscillator circuits do not require any input signal in order to operate.

Oscillators may be divided into two general groups: sine wave oscillators and non-sine wave oscillators. The output of the sine wave oscillator is not quite pure. It is, however, close enough to be considered a pure waveform. In order for the oscillator to work, some of the output signal has to be returned to the control element. This feedback signal must have the proper amplitude and polarity to keep the oscillator running. See Fig. 14-20. Both a transistor and a tube circuit are shown. These are basic circuits.

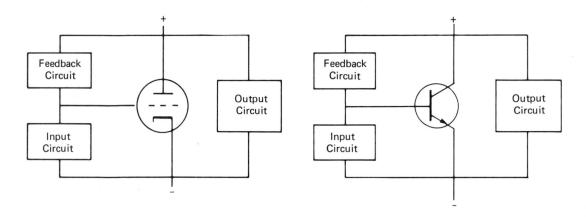

Figure 14-20. Transistor and tube oscillator circuits, with basic block requirements.

There are several ways of classifying oscillators. There are audio frequency, radio frequency, and ultrahigh frequency oscillators. These are classified by their *output frequency*. Another means is by *circuit application*. This group includes local oscillators, beat-frequency oscillators, color oscillators, etc. It is entirely possible that an oscillator could fall into one category in each of these groups. It is probably better to classify oscillators by their circuit features and leave the other groups alone for now. As you look at schematic diagrams of sets and locate the various oscillator circuits, these other categories will become more important. Keep in mind that any oscillator's purpose is to produce an electronic signal.

The RC oscillator. One oscillator circuit found in some consumer electronic products is the RC oscillator. This is diagramed in Fig. 14-21. Two transistors or tubes are required for this circuit. The circuit is biased so that Q_1 conducts more than Q_2. As Q_1 increases its conduction, the collector voltage drops, and a negative voltage is applied through C_1 to the base of Q_2. This, in turn, causes the conduction of Q_2 to decrease. The voltage drop across R_5 also decreases as the transistor conducts. The change in voltage at R_5 is coupled through C_2 to the emitter of Q_1, causing Q_1 to conduct even more. It continues to increase its conduction until it is saturated.

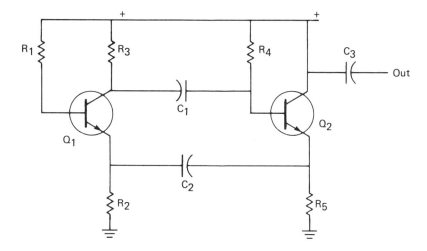

Figure 14-21. A resistance-coupled sine-wave oscillator.

When Q_1 is in saturation, the negative signal is no longer applied to the base of Q_2. Now, Q_2 starts to increase its conduction. A change in emitter voltage at R_5 is coupled through C_2 to the emitter of Q_1. This causes a decrease in conduction in Q_1 and an increase in Q_1 collector voltage. This, in effect, is reflected back to the base of Q_2, through C_1. The effect of conduction by first Q_1 and then Q_2 produces an output signal. The frequency of this circuit is not too predictable because of the variations encountered in component values. It may be controlled by the values of coupling capacitors C_1 and C_2 and by the operating values of the transistors employed in the circuit.

The transformer oscillator. An oscillator using a transformer in its frequency determining section is usually more stable than an RC oscillator. The transformer oscillator circuit is shown in Fig. 14-22. Follow the electron flow in this circuit. Conduction occurs from the negative source, through the emitter, past the base, to the collector, through the primary of

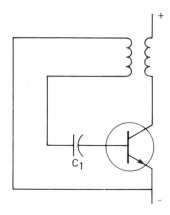

Figure 14-22. A basic transformer-oscillator circuit.

the transformer, and to the positive of the power source. A magnetic field is established in the primary of the transformer as conduction increases. This field produces a voltage in the secondary winding which is coupled through C_1 to the base of the transistor. Increasing the forward bias in this manner makes the transistor conduct more, until it approaches saturation. As the current in the transistor stops changing value, the voltage induced in the secondary of the transformer starts to decrease.

The decrease in secondary voltage effectively starts to reduce the base bias. As the base bias decreases, the conduction in the transistor also decreases. The collector current starts to drop and the voltage produced in the secondary winding of the transformer reverses polarity. Now the transistor conduction decreases until it approaches the cutoff level. When this level is reached the transistor is in a position to repeat the entire cycle.

The LC oscillator. Both the RC oscillator and the transformer oscillator have poor frequency characteristics. Changes in operating conditions of any kind may produce variations in the frequency of the generated signal. A more stable oscillator circuit uses a resonant circuit made up of an inductor and a capacitor. There are several kinds of LC (inductance-capacitance) oscillator circuits in general use. One of these is used as an example of how the LC oscillator works. Refer to Fig. 14-23. This circuit is almost the same as the transformer circuit shown previously. The difference is that a capacitor has been placed in parallel with the base circuit winding of the transformer. The combination of the inductor and the capacitor form a parallel resonant circuit. The resonant frequency is dependent upon the sizes of the two components. Operation is very much the same as for the transformer oscillator circuit except that frequency of oscillation is much more stable.

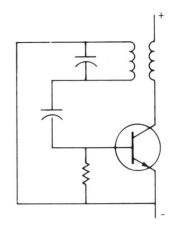

Figure 14-23. The LC oscillator uses a parallel resonant circuit in the base lead to establish frequency.

The crystal oscillator. LC oscillators have found wide acceptance as signal generating devices. The frequency stability of this circuit is very good. There are times, however, when the frequency stability has to be controlled to an extremely close frequency. Such is the case with the color oscillator circuit of the color TV set. In this circuit a crystal oscillator is employed. The quartz crystal acts like a series-resonant LC circuit. When the crystal is used in an electronic circuit, it is made to vibrate mechanically by applying a voltage to it. The mechanical vibrations produce an electrical signal with a constant frequency. A circuit using a crystal is shown in Fig. 14-24. The crystal is connected between the collector and the base in this circuit. It replaces the resonant LC circuit. Capacitors C_1

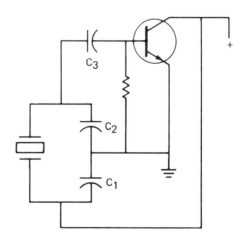

Figure 14-24. The crystal oscillator circuit.

231

and C_2 form a voltage-dividing network between collector and emitter. They determine the amount of feedback in this circuit. The crystal, having a very sharp resonant point, will maintain a close tolerance to this frequency regardless of the changes in circuit values due to temperature or operating conditions.

Non-sine-wave oscillators. The non-sine-wave oscillator, as its name implies, does not produce a sine wave. The output of this oscillator depends upon the design of its wave-shaping circuit. Some of these circuits are free-running. This means that they have a continuous output signal. Other circuits deliver an output only when they are turned on by an input signal. Still others deliver an output that is synchronized to some input triggering signal. Typical waveforms from these circuits include sawtooth waves and square waves.

SUMMARY   Electronic devices used in consumer electronic products require dc voltages. The dc voltages are obtained through the use of a power supply. The power supply converts ac into dc by use of rectifiers. Both semiconductor and vacuum tube rectifiers operate on the same principle. Conduction occurs when the cathode is negative with respect to the anode or plate.

The output of the rectifier is either half of the input wave or the full input wave, but with a specific polarity. The pulsating dc wave is smoothed by use of a filter circuit. The filter eliminates any pulsations, or ripple, in the waveform.

In some instances, it is desirable to obtain a higher output voltage without the use of a transformer. Often, a voltage multiplying circuit is used for this purpose. Capacitors are charged during part of the conduction cycle. The charged capacitor adds to the applied voltage of the circuit when it discharges. This provides a higher output voltage to the load. Voltage multipliers are used to provide two to five times the input voltage, as required by the circuit.

It is possible to use a source other than the power line for an ac source. Some currently produced TV sets are using the 15,750-Hz horizontal signal as a source. This *scan derived* power source is used as a secondary power source to provide the required operating voltages in the set.

Rectifiers are also used as detectors. The halfwave diode circuit is used to detect AM radio signals. TV video detectors are also halfwave rectifier circuits. FM detectors use a frequency detection circuit made up of two diodes.

Voltage regulators are used to provide a steady dc output voltage in the set. Zener diodes, transistors, and vacuum tubes are used in these regulator circuits.

The amplifier is a device that uses a small amount of power to control a larger amount of power. The circuit used consists of an input circuit and an output circuit. The input circuit is the control circuit. The output circuit is the controlled circuit. Junction transistors, FETs and tubes are all used as amplifiers. The control circuit actually changes the resistance in the device. The output circuit consists of a load resistance in series with the resistance of the transistor or tube. These two components form a voltage-dividing network. Variations in the resistance of the active device produce changes in output voltage. These changes produce a varying dc voltage which looks like and acts like the input signal. In most circuits the output wave is larger than the input signal. The term *amplifier* is used because of this action.

Amplifier circuit classifications are based upon the common element in the circuit. These are: the common emitter, source, or cathode; the common base, gate, or control grid; and the common collector, drain, or plate. The common emitter-source-cathode circuit has the highest gain of the three configurations.

Amplifiers are classifed as class A, class AB, class B, or class C. This classification refers to their mode of operation. The class A amplifier is on 100% of the time. It faithfully reproduces the input signal at its output element. The class B amplifier is on 50% of the time. It is used as a power amplifier. Its output waveform is similar to that of a halfwave rectifier. Class AB amplifiers are a compromise between class A and class B amplifiers. They are used to minimize distortion and to provide power amplification in the circuit. The class C amplifier is on less than 50% of the time. It is used as an RF amplifier. A resonant circuit is used with the class C amplifier in order to reconstruct the original wave shape.

An oscillator circuit is used to generate a wave. Most oscillator circuits use the amplifier configuration plus a feedback circuit in order to generate a wave at a specific frequency. The frequency is dependent upon the resonant circuit and the feedback components.

The oscillator used in tuners of radios and TV sets produces a sine wave signal. The oscillator used in vertical and horizontal circuits in TV sets produces a sawtooth waveform. Other oscillator circuits are used to produce square waves. Shape and frequency of the oscillator wave depends upon the specific circuit used.

QUESTIONS

1. Draw a schematic diagram of a semiconductor halfwave-rectifier circuit with a negative dc output.

2. Draw the output waveform for the above circuit.

3. Draw a schematic diagram of a fullwave rectifier circuit. Show electron flow during the complete input cycle.

4. Explain the term *ripple voltage*.

5. Why are active power filters used in electronic devices?

6. What does a voltage regulator use to control the level of output voltage?

7. Draw a schematic of a common-emitter amplifier circuit. Show input and output connection points.

8. Draw the input and output waveform for the circuit in Ques. 7.

9. Which component in a voltage-doubler circuit is used to develop the double input voltage?

10. Which amplifier circuit is used to provide high power output for a loudspeaker?

11. What are the differences between an amplifier and an oscillator?

12. What determines oscillator frequency?

13. What are the wave shapes obtained from the oscillator?

14. Where are these wave shapes used in the TV set?

Section Five

Circuit
Analysis

Audio Device Analysis

15

Earlier in this book you were exposed to block diagrams of consumer electronic products. These were for the audio amplifier, AM and FM radio, and both black-and-white and color TV. Each block function was explained. Succeeding chapters covered signal paths through the various blocks and devices. Now, after having discussed active and passive devices, as well as circuit analysis, the next step is to look at the signals and operating voltages in these amplifiers, radios, and TV sets as they apply to actual components.

There are many amplifier circuits found in home entertainment sets. All of these have but one purpose. This is to increase the signal level to a point where it can be used to operate a loudspeaker. This applies for monaural and for two- and four-channel stereo systems. The audio section of a stereo system is composed of two or four identical amplifiers. Each one operates a speaker. The circuitry used is the same for each. Signal processing is often the same. The results of an examination of one channel of the amplifier can be applied to other amplifiers in a system to show how each works. The discussion of audio amplifiers in this chapter will use a single-channel amplifier as an example.

The block diagram of an audio amplifier is shown in Fig. 15-1. A low level of signal is at the input. This signal is amplified greatly in the preamplifier and voltage amplifier blocks. The power amplifier converts the high signal voltage into a useful power level in order to operate the output transducer, or speaker. Sine waves are shown on the block dia-

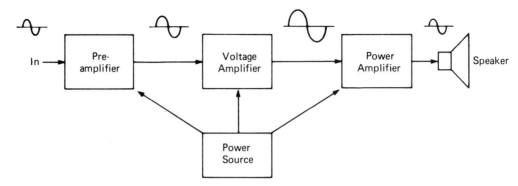

Figure 15-1. An audio-amplifier block diagram, showing signal paths and signal amplitide.

gram which represent signal voltage levels through the amplifier. The power source is connected to the various transistor or tube circuits in order that amplification may occur.

Now look at the schematic diagram shown in Fig. 15-2. This illustrates an amplifier which has all of the previously mentioned blocks. Each block is identified on the actual schematic with a heavy line. It is sometimes difficult to identify the exact point where one block ends and another starts. There often is overlap between blocks in an actual set. Let's start with the power supply. This is a transformer operated circuit with a 120-volt input. The secondary of the transformer steps down the voltage to an operating value of about 25 volts. A fullwave bridge-rectifier is employed in order to change the ac into pulsating dc. A very high value capacitor is used as the filter in the power supply.

The output of the power supply is connected to a line which, in this circuit, runs along the top of the schematic from the V+ terminal, through R_{21}, to the junction of R_5 and C_3. The common connection is another line, however this line is near the bottom of the diagram. It runs from the common symbol, on the right, to the outside connection on the input jack. Several voltage-divider networks are used in this circuit. One consists of Q_1 and R_5. Another is R_{14}, Q_3, and R_{13}. A third is R_{22} and Q_5. One other is Q_7, R_{26}, and R_{25}. The last one consists of Q_{10} and Q_9. These voltage dividers are employed in order to establish operating conditions for each transistor. A complete service schematic will include operating voltages at each transistor element.

The voltage-divider circuits just identified are for the output circuits. There is another set of voltage dividers used at the control element of each transistor stage. An example of one of the bias circuits is the network made up of R_{14}, Q_3, and R_5. This establishes the operating point for Q_3.

All of the transistors except Q_9 and Q_{10} are operating as Class A amplifiers. Q_9 and Q_{10} are Class B. Look at Fig. 15-3. This is the same circuit. Now the signal path is identified. A sine wave signal is presented at input jack A. It is coupled through C_1 to the base of Q_1. Variations in signal level at the base cause the dc voltage at the collector of Q_1 to vary.

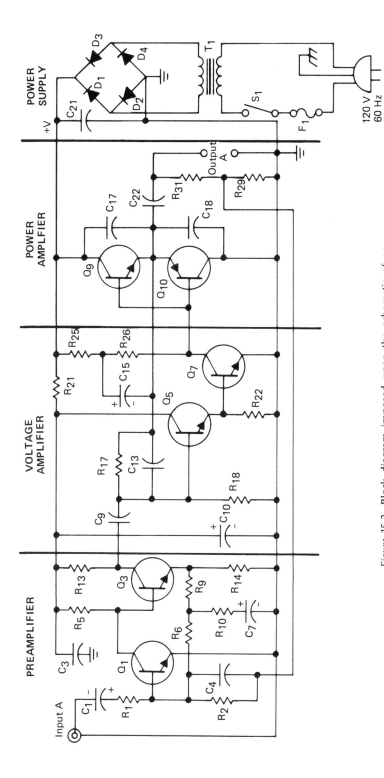

Figure 15-2. Block diagram imposed upon the schematic of an amplifier.

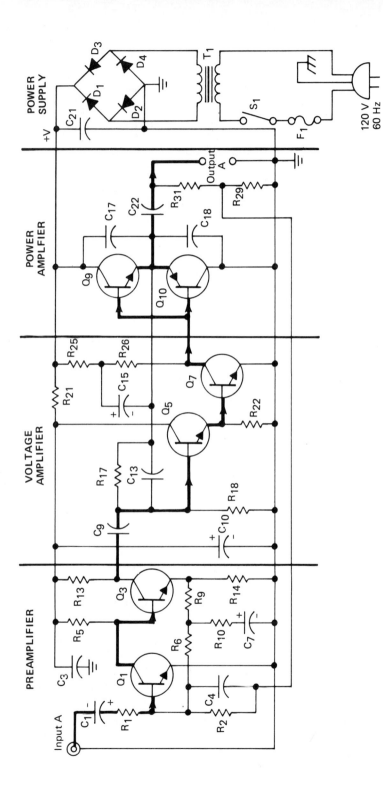

Figure 15-3. Signal path through the amplifier.

This changing dc voltage is directly coupled, or wired, to the base of Q_3. These variations, in turn, cause changes in the collector voltage of Q_3. Now the variations are coupled through capacitor C_9 to the base of Q_5. Using a capacitor as a coupling device removes the dc values from the signal. This means that the dc voltage, which may be varying from 8 volts to 10 volts, does not appear at the base of Q_5. Only the 2-volt variation is coupled to this base.

Both Q_1 and Q_3 are common emitter circuits. These circuits offer a high voltage gain in each stage. Q_5 is operated as a common collector amplifier. This means that the output signal is taken from the emitter element. It is directly coupled to the base of Q_7. Q_7 is operated in a common emitter circuit. Its output is directly coupled to the bases of Q_9 and Q_{10}.

All transistors except Q_{10} are NPN types. Q_{10} is a PNP transistor. Its operating polarity is the opposite of that of the NPN transistors. This is used to advantage in the output stage of the amplifier. Both Q_9 and Q_{10} operate as class B amplifiers. They are biased at the cutoff point. Conduction will occur in the transistor when the input signal causes the transistor to turn on. The NPN transistor, Q_9, will only conduct during the positive half of the input wave cycle. The PNP transistor, Q_{10}, will only conduct during the negative half of the input wave cycle. The output circuit is at the emitters of these two transistors. The output is coupled through C_{22} to the speaker. This circuit is a common collector circuit. It provides the low-impedance, high-current output required to drive the speaker. The advantage of using a class B operating mode for the output stages is that a higher power output is obtained. Each output transistor operates during half of the complete cycle. These transistors can operate at higher than normal power values because they are off and cooling during the other half of their cycle.

The amplifier schematic used as an illustration is typical of a good number of solid-state amplifiers. Signal flow is normally from left to right on the schematic. The exact path for the signal is usually easy to identify. If one works backwards through the schematic, going from output to input, the signal path will become evident. In this example, the signal at the output terminal is connected to the emitter of the output transistors, and the output transistors bases are connected to the emitter of Q_7. This is one method of identifying the signal path of the unit being examined. It works with both transistor and tube circuits.

Figure 15-4 illustrates both the transistor and the tube amplifier. They are quite similar in construction and in operation. These are examples of capacitive-coupled voltage amplifiers. The input signal is coupled through C_1 to the control element of the first transistor or tube. These circuits are common emitter and common cathode. The output elements of this stage are the collector and the plate. Capacitor C_2 is a coupling capacitor used to transfer the signal to the second-stage control elements. These, as in the other stage, are the base and control grid. These stages are also common emitter and common cathode circuits. The output from the

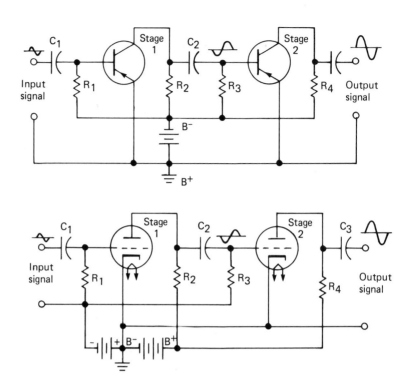

Figure 15-4. Comparison of transistor and tube amplifier.

second stage of each amplifier is coupled through capacitor C_3 to whatever stage is next in line.

Follow the signal waveform through each circuit. The input signal is comparatively small. It starts with a positive half cycle. The first-stage output signal is amplified and has changed polarity by 180 degrees. This is typical in a common-emitter or common-cathode amplifier circuit. The output waveform, amplitude, and polarity are the same as the input to the next stage. The output signal is amplified again in the second stage. Its phase is reversed 180 degrees, also. The exact amount of amplification depends upon the circuit and the active device used in the circuit. These factors are determined by the design engineers.

It is possible to have gains of 50 to 300, or more, in one transistor amplifier stage. Tube amplifier stages normally have gains of less than 100. Gain is very easy to determine. The formulas for voltage gain and power gain are:

$$\text{Voltage gain} = \frac{\text{Output signal voltage}}{\text{Input signal voltage}}$$

and

$$\text{Power gain} = \frac{\text{Output signal power}}{\text{Input signal power}}$$

These formulas are used for either a single stage or a complete amplifier.

15-1
Output Circuits

Several circuits are used as power output stages in amplifiers. Three of these output circuits are: the single-ended transformer coupled, the push-pull transformer coupled, and the complementary-symmetry transformerless. The schematic diagram for each circuit is shown in Fig. 15-5. Since current state-of-the-art emphasis is on semiconductor circuits,

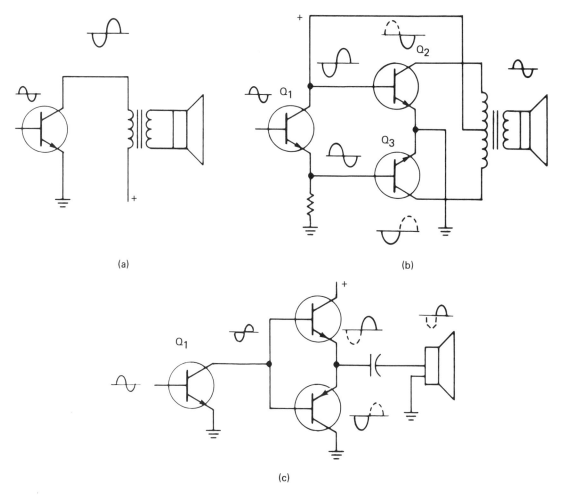

(a)

(b)

(c)

Figure 15-5. Common output circuits: (a) single-ended; (b) push-pull; and (c) complementary symmetry.

these are the type illustrated. Vacuum tube circuits, as have been illustrated previously, operate in similar fashion. Circuit (a), of Fig. 15-5, is the *single-ended transformer-coupled* circuit. Electron flow is controlled by the emitter-base circuit. The transistor emitter-collector and the primary of the transformer form a voltage-dividing network. The transformer primary is the load in this circuit. Current changes in the transistor-transformer circuit cause the magnetic field in the primary of the transformer to vary. This produces a voltage in the secondary winding. The voltage induced in the secondary winding causes a current in the circuit made up of the secondary winding and the speaker. Audio output transformers are usually voltage step-down transformers. Applying the law stating that power out equals power in, we know that if the secondary voltage is low, then the secondary current increases. The increase in current in the secondary is necessary for speaker operation, since the speaker is a current-operated device.

The second circuit illustrated in Fig. 15-5 is a *push-pull* circuit. It uses an output transformer that has a center tapped primary and a single secondary. One half of the primary of the output transformer is connected to the collector of Q_2. The other half if connected to the collector of Q_3. The center tap of the transformer is connected to the positive voltage source. The current path is from negative, through the emitter of Q_2, to the collector of Q_2, through half of the primary, and then to the positive power source. Q_3 current takes a similar path, using the emitter-collector of Q_3 and the other half of the transformer primary. There is only one secondary winding in this transformer. Any current moving in half of the primary induces a voltage in the secondary. This means that, when one transistor is conducting, there will be an output signal at the speaker. When the second transistor conducts, there will also be an output at the speaker. Now, follow the signal through the circuit from the input to the output. Q_1 is called a *phase splitter* circuit. It has one input and two outputs. It is connected with an output at the collector which produces a signal whose phase is 180 degrees away from the input signal. It also has an output connection at the emitter, and this output is in phase with the input signal.

These two output signals are fed to the bases of Q_2 and Q_3. Both transistors are biased to operate in, or nearly in, a class B manner. This means that they are biased at the point of cutoff. The input signals turn them on at different times. When Q_2 is conducting, Q_3 is off. When Q_3 is on, then Q_2 is off. This permits higher power operation than if both were on all of the time. At the output of each transistor, the waveform produced looks like that of a halfwave rectifier. Using a split primary and single secondary allows the waveform to be recreated in the secondary, forming the original shape.

Circuit (c) of Fig. 15-5 is called a *complementary-symmetry transformerless* output circuit. The output transistor circuit uses one NPN and one PNP transistor. They are series connected to form a voltage-divider circuit. It is not necessary to use a phase splitter because the polarities of the two transistors are opposite. These, too, are biased at the cutoff point. The incoming signal turns on first one and then the other transistor. Each conducts during half of the cycle of the input signal. The output is capacitive coupled to the speaker. Slight changes in the value of emitter voltage at the point where both emitters are connected to each other produce a signal wave at the capacitor. This signal is coupled to the speaker through a large value of capacitance in order to operate the speaker with sufficient power.

15-2
IC Amplifiers

One method of simplifying production and servicing in the past few years has been to use an *integrated circuit* instead of discrete components. An early use for the IC was in audio circuits. This was probably because of the lower frequencies involved. Since their introduction, the use of ICs has increased so greatly that these *chips* are now found in most sections of amplifiers, radios, and TV sets. Very little can be done with the IC, by the technician, as far as tracing the signal path is concerned. The amplitude and the phase of the signal at the output depend on the internal design of the chip. Symbols for the IC and a schematic diagram of an IC amplifier are shown in Fig. 15-6. The only easy way of checking signal flow is to use a schematic diagram having waveform information. The amplitude and shape of the signal at the IC input and output are compared with those on the service literature. If they look alike, then the IC is probably good. If the output signal does not agree with that shown on the service manual and the input signal looks good, then the IC should probably be replaced.

15-3
Tape Recorders

Amplifiers used in tape playback-and-record units are similar in design to those found in other audio devices. In the playback mode of operation, signals follow the same paths through most of the audio circuits. The signal is picked up from the tape head and fed into the first preamplifier stage. From there it moves through the amplifier, stage by stage, until it reaches the output. The output is changed into sound waves by the speaker.

An additional circuit is used when a tape recording is made: the bias oscillator. *Bias* is defined as a voltage which sets up the operating conditions for the transistor or tube. An *oscillator* is an electronic-signal producing device. A *bias oscillator* is a device that produces an electronic signal at a specific frequency which is used to set up the operating conditions.

The bias oscillator is used for two purposes in a tape recorder. One purpose is to "clean" the tape before it is recorded. An erase head is employed for this purpose. The signal from the bias oscillator is fed to the erase head. The frequency of this signal is in the range of 35 to 100 kilohertz. This frequency is such that any information on the tape is *erased,* or removed electronically. The particles of oxide on the tape are then ready to be arranged by electronic means to represent the recorded information.

The second purpose of the bias oscillator signal is to help make a distortion-free recording. When a coil of wire is made into a magnet and the current applied, there is a slight time lag while the current increases.

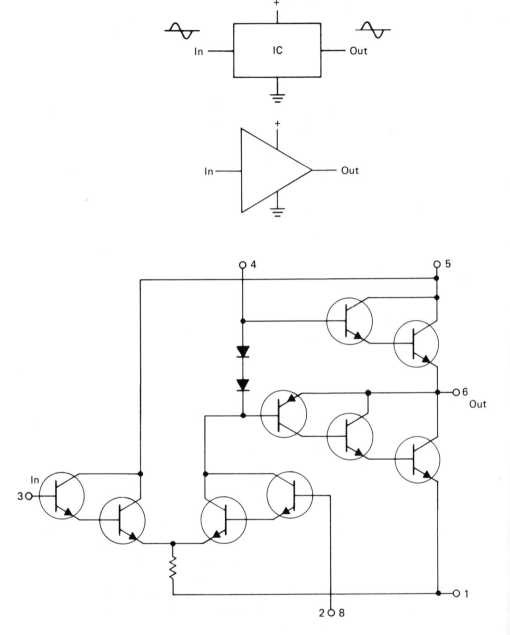

Figure 15-6. Schematic symbols for integrated-circuit amplifiers, and the actual circuit inside one IC.

This disparity also occurs when the magnet is turned off. The bias oscillator signal is added to the audio-frequency information signal in order to minimize this effect. Figure 15-7 shows the complete amplifier circuit, with the bias oscillator located at the lower left of the diagram. Note the

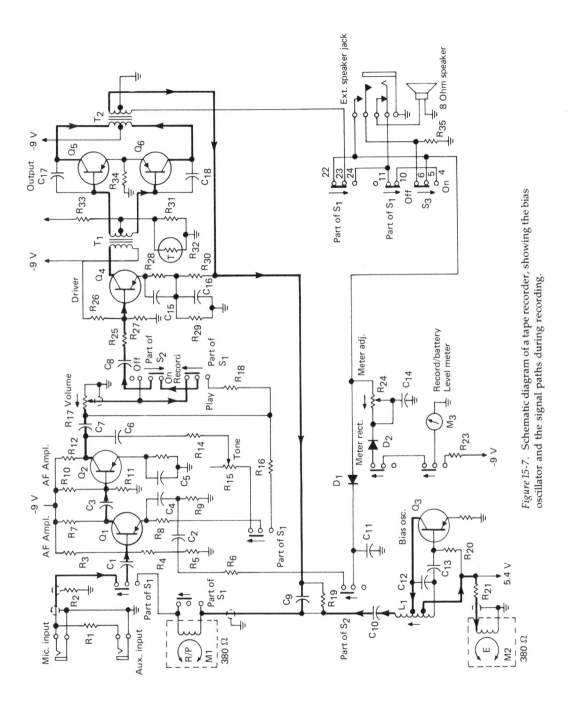

Figure 15-7. Schematic diagram of a tape recorder, showing the bias oscillator and the signal paths during recording.

247

two outputs from the oscillator. One is connected to the erase head, E, through R_{21}. The second output is coupled through C_{10} to the record-playback head, R/P. The signal path for the signal when the unit is in the record mode is shown by a heavy black line. It starts at the microphone (Mic) input and flows through the amplifier to a point just before the power output stage. At this point it is fed to the record-playback head and the information is recorded by the tape head. The information signal and the bias oscillator signal are combined and used in this manner during the production of a recorded tape.

SUMMARY   The audio amplifier processes a signal whose frequency is in the audio range of 20 hertz to 20 kilohertz. The input to the audio amplifier is usually a very low amplitude signal. This input signal is increased to several hundred times its original value by use of transistor or tube amplifier-stages. The signal is amplified through a series of voltage amplifiers. It is then sent to the output stage. The output stage is a power amplifier stage. It is necessary to develop power instead of voltage in order to operate the speaker.

The schematic diagram of an amplifier may be divided into blocks. These are the same blocks as those identified in the block diagram. The signal amplitude, shape, and polarity may be shown on the schematic diagram. Technicians use this information when troubleshooting in order to locate a weak or nonoperating stage.

Two types of voltages are indicated on most schematic diagrams. One is the dc operating voltage required to establish the proper biases for operation. This type of voltage is provided by the power supply. The other type is the signal voltage, and it is furnished by the input transducer. This is the intelligence that is amplified and sent to the speaker where it is converted into sound waves.

Each transistor or tube is connected as part of a voltage-dividing network in the circuit. Changes in internal resistance in the transistor or tube are used to develop an output signal. The output signal is then coupled to the next stage.

Gain is equal to the comparative amount of amplification in a stage or complete unit. The output is compared to the input: gain = output/input. This is true for voltage (signal) and power. The numerical solution to the equation represents the amount of amplification in the specific stage or circuit.

Several different output circuits are used in order to develop power in the amplifier. Three common circuits are the single-ended, push-pull, and complementary symmetry. These circuits are used in audio circuits. They are also found in vertical circuits in TV sets. The purpose of the output stage is to develop current for the output transducers. These circuits also serve as impedance matching devices in order to transfer maximum power from the stage to the output device.

Integrated circuits are used in many different ways in electronic devices. One typical way is as an amplifier. The IC is employed as a voltage amplifier and, in some cases, as a complete voltage and power amplifier. The actual electronic circuit used in the IC depends upon the design of the unit.

Tape recorders use audio amplifiers. In the playback mode the tape recorder is used to convert the signals on the tape into sounds one is able to hear. Electronically, it is an amplifier. In the record mode, another circuit is added. This is the bias oscillator. The bias oscillator serves two purposes. One is to erase any information on the tape. The other purpose is to compensate for magnetic fields in the recording head in order to prevent distortion in the recording. Bias oscillator frequencies range from 35 kilohertz to 100 kilohertz.

QUESTIONS

1. Name the blocks found in an audio amplifier.

2. Describe the function of each block in an amplifier.

3. Using the schematic shown in Fig. 15-2, identify the kinds of coupling found between each transistor in the amplifier.

4. Why is it necessary to have a dc voltage at the output element of the transistor or tube that is not equal to the supply voltage?

5. Why are output stages called power amplifiers?

6. What is meant by the term *push-pull*?

7. A stage has an input signal of 0.2 volt. The output signal is 1.5 volts. What is the voltage gain of this stage?

8. How is an IC amplifier tested?

9. What are the two purposes of the bias oscillator in a tape recorder?

10. What are the output connections of the bias oscillator in the tape recorder?

Radio Circuit Analysis

16

In this consideration of actual circuit operation in working sets, we will first discuss the block diagram for a typical set. Next, we will look at a schematic diagram which identifies the basic blocks on the schematic. Finally, operating voltage circuits, signals, and signal path circuits are discussed. In some cases there is an overlap due to the fact that the same block diagram is used for more than one specific set.

The AM and the FM radio units have the same basic block diagram. This is illustrated in Fig. 16-1. The modulated RF signal is received and

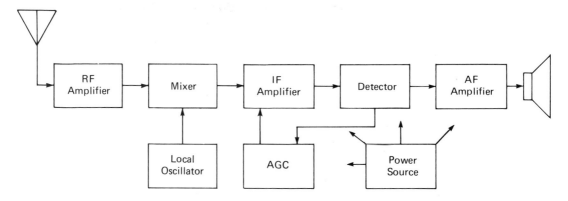

Figure 16-1. Block diagram of an AM or FM radio.

amplified in the RF amplifier. From this block it is sent to the mixer. Some sets omit the RF amplifier and inject the RF signal from the antenna into the mixer. The local oscillator block produces its carrier signal which is sent to the mixer. The mixer output is a modulated carrier at the IF frequency. This signal is amplified in the IF amplifier and sent to the detector where the information is removed from the IF carrier. The information is then processed through the audio amplifier and sent to the speaker.

The differences between an AM and an FM radio appear in the frequencies as they are received, the IF frequency, and the means of detection. AM radios for home entertainment operate in a frequency band from 540 kilohertz to 1.6 megahertz. The IF frequency is 455 kilohertz. Diode halfwave detection is employed. The FM receiver uses the group of frequencies from 88 megahertz to 108 megahertz. Its IF frequency is 10.7 megahertz. The detection system is frequency sensitive and may use a discriminator or ratio detector circuit.

Audio circuits for both kinds of radios are similar. The FM stereo receiver uses a two-channel system. Four-channel stereo receivers use four audio amplifiers and speaker systems. AM and FM monophonic systems use a single-channel audio amplifier. Power supply circuits are also similar. Here, the biggest difference, as evidenced by the amount of power required to operate the radio as well as by the purity of the dc voltages used, is in the amount of filtering required for operation of the set.

16-1
AM
Radio Receivers

Figure 16-2 illustrates the blocks found in several makes of the lower priced AM radios. This type of radio does not use an RF amplifier. An additional tuned stage is employed when the receiver does have an RF amplifier, and a circuit of this type is illustrated later in this chapter. In most instances, the output of one block and the input to the next block overlap. The separation between blocks occurs at a point somewhere near the coupling device used. In the radio, it is possible to presume that this point is near or in the IF transformers in that section.

Let's look at the dc operating-voltage circuits used in this set. The dc operating voltage is obtained from the power supply. The output point of the power supply is the junction of D_2 and C_{22}. The power output-circuit current goes from the power source, through R_{17}, through the emitter-collector circuit of Q_5, through the primary of T_4, and back to the power source. Electron flow through Q_4 takes this path: circuit common, R_{16}, R_{15}, Q_4 collector-emitter, R_{14}, and power source. Q_2 electron flow is: circuit common, primary of T_2, Q_2 collector-emitter, R_7, R_{18}, and power source positive. Other circuit voltage-dividers use the same kind of circuitry.

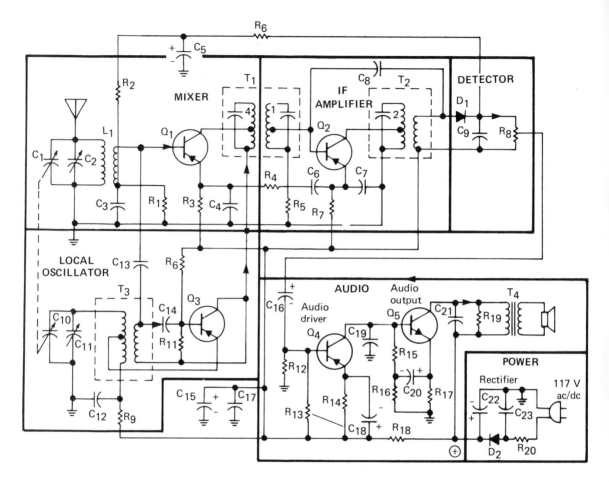

Figure 16-2. Block diagram of an AM radio, showing the blocks as they relate to the schematic.

Follow the signal through the radio schematic, as shown in Fig. 16-3. The output of the oscillator Q_3 is connected to the base of Q_1 by means of capacitor C_{13}. This provides two signals at the base of this transistor. These two signals are heterodyned in the mixer transistor Q_1. The output of the mixer transistor consists of four signals. These are: the RF signal from the antenna, the signal from the local oscillator, another signal whose frequency is the sum of the two input signals, and a signal whose frequency is the difference between the two input signal frequencies. Transformer T_1 is tuned to 455 kilohertz. The dashed lines connecting C_1 and C_{10} indicate that these two variable capacitors are connected to the same tuning shaft. They both turn when the turning knob on the set is rotated. C_2 in this circuit is used as an alignment adjustment. The received signal is coupled through the secondary of L_1 to the base of Q_1. It is amplified in the output circuit of Q_1 and the primary of T_1.

252

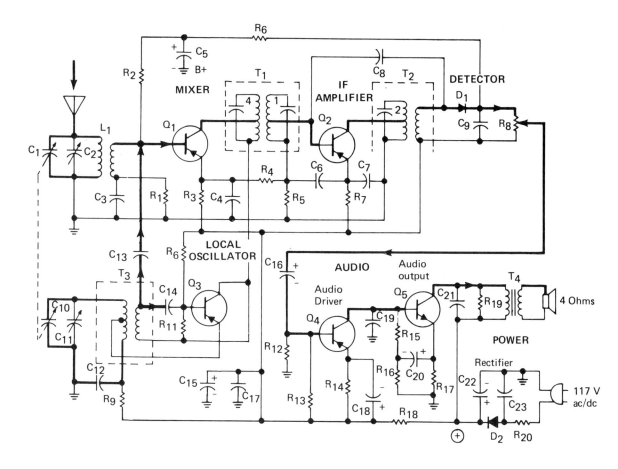

Figure 16-3. Signal paths for the AM radio. Note the two signals sent to the mixer.

At the same time this is happening, the local oscillator is producing a non-modulated carrier wave. The frequency of this wave is determined by a resonant circuit which includes C_{10}. C_{11}, C_{12}, and the primary of T_3, the oscillator coil. The output of the oscillator Q_3 is connected to the primary of L_1, also. This provides two signals at the secondary of L_1. These two signals are heterodyned. Transformer T_1 is tuned to a frequency of 455 kilohertz. Because of this tuning, only one of the four signals produced at the mixer is coupled into the IF amplifier by T_1. This signal is an amplitude modulated signal whose carrier frequency is 455 kilohertz.

The IF signal is coupled to the base of Q_2 by the secondary of T_1. After amplification in the emitter-collector and primary of T_2, it is coupled to the detector diode D_1 by the secondary of T_2.

The circuit made up of the secondary of T_2, D_1, and R_8 is a halfwave rectifier circuit. The IF signal is rectified and the result is a signal whose outline is that of the modulation. The carrier is filtered out by the action of

C_9. The audio signal that remains is sent to the base of the audio driver through coupling capacitor C_{16}. R_8 is the volume control. It works in a manner similar to a variable voltage-divider network. The position of the arm (indicated by the arrow) in relation to the cathode of D_1 determines the amount of signal that is coupled to the audio amplifier stages. The audio signal path is shown in Fig. 16-3, also. (Audio paths are covered in Chap. 15.)

16-2
FM
Radio Receivers

The portion of the FM radio illustrated in Fig. 16-4 is similar in operation to the AM radio just described. There are some differences, however, which should be discussed. This radio has an RF amplifier stage in front of the mixer. This stage provides additional signal-gain in the set. It also increases the ratio of signal to noise received. A high signal-to-noise ratio in a receiver decreases the annoying background hiss produced by atmospheric discharges. Another feature of this particular radio is that it has three stages of IF amplification. This also produces a larger amount of signal for the audio amplifier. In addition, it increases the ability of the radio to receive and process weak signals from distant broadcast stations.

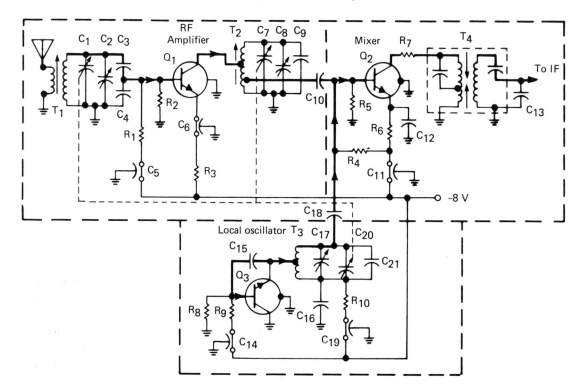

Figure 16-4. Schematic diagram of the tuner section of an FM radio. Blocks are labeled. Signal paths are indicated by the heavy lines with arrows.

Follow the signal through the radio on the schematic diagrams shown in Figs. 16-4 and 16-5. The RF signal is selected by the resonant circuit C_1, C_2, C_3, C_4 and the secondary of T_1. This radio uses a three-section tuning capacitor, as shown by the dashed lines connecting C_1, C_7, and C_{17}. The input and the output circuits of the RF amplifier are both tuned. The oscillator circuit is tuned, also. C_1, C_7, and C_{20} are the *ganged variable-capacitors* used in the radio.

The RF signal is amplified by Q_1 and then fed to the resonant circuit T_2, C_7, C_8, and C_9. The signal is coupled from this circuit to the base of Q_2 by C_{10}. Q_2 is the mixer transistor. It requires two input signals in order to produce the 10.7-megahertz IF signal.

The local oscillator, Q_3, generates a carrier signal which is 10.7 megahertz higher than the signal frequency tuned in the RF amplifier. This signal is also fed to the base of the mixer transistor, Q_2. It is coupled from the oscillator resonant circuit T_3, C_{17}, C_{20}, and C_{21} by capacitor C_{18} to the mixer stage. The output of the mixer stage is a resonant circuit made up of the primary of T_4 and the capacitor in parallel with this winding. This transformer is tuned to 10.7 megahertz. The IF carrier and its modulation are coupled to the secondary of this transformer and fed to the base of the first IF transistor, Q_4 (Fig. 16-5). Another tuned circuit is located in the output portion of this transistor. It consists of the primary of T_5, R_4, and the collector-emitter circuit of Q_4. This transformer is also resonant at a frequency of 10.7 megahertz. The signal is coupled to the base of Q_5 by the secondary of T_5. Both the first and second IF amplifiers work in this manner.

The output of the third IF amplifier is coupled through T_7 to the detector circuit. This circuit is called a *ratio detector*. It is a type of frequency detector that eliminates the need for a limiter stage in the receiver. Slight variations in frequency from the 10.7-megahertz IF frequency cause one of two diodes, D_2 or D_3, to conduct. This, in turn, develops a voltage across C_{15}. This voltage represents the modulation information placed on the carrier at the transmitter. The output signal is connected at the center tap of the secondary winding through R_{13}. This signal then goes to a filter network consisting of C_{10}, R_{15}, and C_{14}. This network removes any IF carrier signal that might remain. The output of this filter network is fed to the input of an audio amplifier through a volume control circuit.

There is another output shown in this schematic. This is the stereo trigger-circuit output. It feeds the stereo decoder section of the radio and turns on the stereo decoder whenever this trigger signal is received.

16-3
FM
Stereo Receivers

Much information is contained in the composite FM-stereo signal. This signal is illustrated in Fig. 16-6. Analyzing the broadcast signal reveals several discrete signals on the carrier. The group of frequencies from zero to 15 kilohertz make up the monaural FM signal. This is called the $L + R$ signal. The small signal tone at 19 kilohertz is the pilot signal

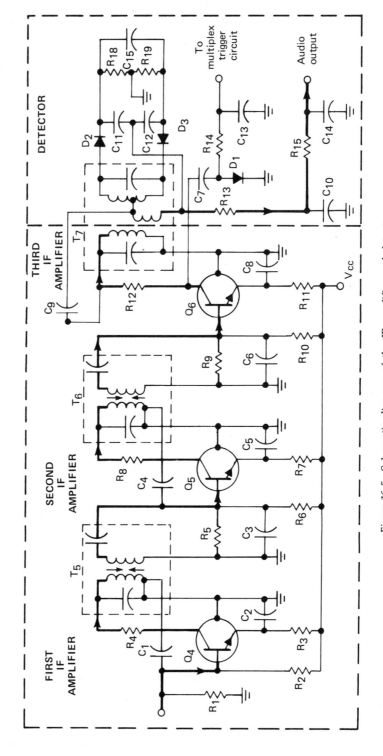

Figure 16-5. Schematic diagram of the IF amplifier and detector sections of an FM radio. Signal paths are shown by heavy lines with arrows.

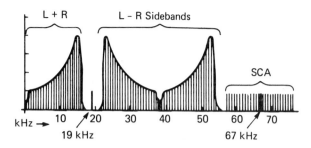

Figure 16-6. The signals present in the FM station carrier. Bandwidth of the carrier is 75 kilohertz in order to carry all of this information.

subcarrier. It is used to synchronize the carrier signal in the receiver with the broadcast signal. The stereo information is added to the main carrier as a double-sideband suppressed-carrier signal. This suppressed carrier is located at 38 kilohertz. These sideband signals are $L - R$ signals. One is an $L - R$ upper-sideband signal and the other is an $L - R$ lower-sideband signal. The $L - R$ signals and the 19-kilohertz pilot subcarrier make up the stereo portion of the FM signal.

It is possible to have one additional signal on the main FM carrier. This signal is called the SCA (for *subsidiary carrier assignment*) signal. The SCA signal is not broadcast by all stations. It is used primarily for background music systems. The frequency range of this signal is centered at 67 kilohertz on the carrier. Special receiver circuitry is required in order to receive this signal.

The block diagram of a stereo decoder system is illustrated in Fig. 16-7. This represents one type of system used in FM stereo receivers. Follow the composite stereo signal through these blocks. The signals have already had the 10.7-megahertz IF carrier removed by the action in the detector.

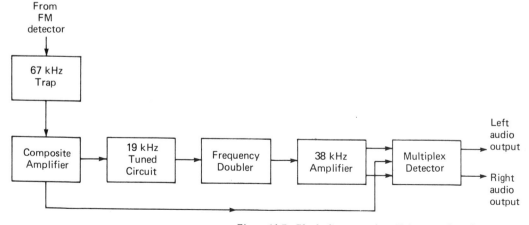

Figure 16-7. Block diagram of an FM stereo-decoder system.

The first block is a 67-kilohertz *trap*. This tuned circuit rejects, or stops, any frequencies around 67 kilohertz. If an SCA signal is present it is blocked at this point. Lower frequencies are allowed to pass by the trap due to its design. A filter of this type is called a trap because it blocks a specific frequency and permits all other frequencies to pass.

The amplifier block have one of several names, such as the *bandpass amplifier,* or the *composite-signal amplifier.* Its purpose is to amplify the mono and stereo signals as they come from the detector. This block has two outputs. One feeds the 19-kilohertz amplifier and frequency-doubler circuit. The other bypasses these blocks and goes directly to the detector block.

The purpose of the 38-kilohertz amplifier is to provide a carrier in order to demodulate the $L - R$ information. These signals are transmitted as double-sideband suppressed-carrier signals on the FM station carrier. The 38-kilohertz subcarrier has to be regenerated in order to detect this signal. A 19-kilohertz tone is generated and added to the carrier at the transmitter. This frequency is half of the 38-kilohertz carrier needed for demodulation of the signal. The 19-kilohertz signal is amplified and then its second *harmonic* is used and amplified in order to replace the suppressed 38-kilohertz carrier for the $L - R$ information.

The output of the 38-kilohertz amplifier is sent to the stereo detector block. The two signals, $L + R$ and $L - R$, are mixed, or added, in this detector block. The result of this action is one left-channel signal and one right-channel signal. These signals are then fed to the left- and right-channel audio amplifiers.

The schematic of a stereo decoder is illustrated in Fig. 16-8. The blocks are identified at the top of the drawing. Follow the signal paths through this schematic. The composite stereo signal is fed from the receiver detector system to the input terminal of the decoder unit. This signal first passes through L_1 and C_1, which are used as a 67-kilohertz trap in order to block any SCA signal. It is then coupled through C_2 to the base of Q_1. The composite stereo signal is fed from the emitter of Q_1, through C_4, to the center-tapped secondary of the detector transformer T_3. The collector circuit of Q_1 is used to amplify the 19-kilohertz pilot signal. This signal is further amplified by Q_2 and then sent to a frequency doubling circuit. The frequency doubler uses the secondary of T_2 and diodes D_1 and D_2 in a fullwave-rectifier type circuit. This circuit produces pulses of the input signal at two times the input signal frequency ($2 \times 19 = 38$).

The 38-kilohertz carrier signal is fed to the primary of T_3 through the amplifying action of the Q_3 collector circuit. The secondary of T_3 is connected to two sets of detector diodes. These sets are D_3 and D_4, which provide the left audio signal, and D_5 and D_6, which provide the right-channel audio signal.

The stereo signal is re-created as both the $L + R$ signal and the $L - R$ signal mix in the secondary of T_3. Diode D_3 conducts on the negative alternation of the $L + R$ signal. Diode D_4 conducts on the positive alter-

258

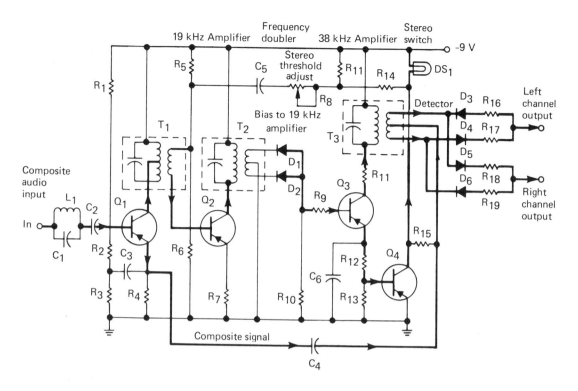

Figure 16-8. Schematic diagram of an FM stereo decoder. Signal paths are shown by heavy lines with arrows.

nation of the $L + R$ signal. There is but one output. Therefore, the entire $L + R$ signal is re-created at the output. During a stereo broadcast, the $L-R$ signal is added to the $L + R$ signal as the detector diodes D_3 and D_4. The result of adding $L + R$ and $L - R$ produces a 2-L, zero-R signal at this point. This is the left channel signal.

Diodes D_5 and D_6 are connected in opposite polarity to the same points as diodes D_3 and D_4. Their output is equal to an opposite polarity signal. This signal is called the $-(L - R)$ signal. This is the same as a $-L + R$ signal when simplified. When the $-L + R$ signal is added to the $L + R$ signal, the result is a zero-L + 2-R signal. This makes up the right channel audio signal.

The remaining circuit consists of Q_4 and the stereo indicator lamp, DS_1. This is the circuit used to turn on a stereo-indicator lamp. It only works during a stereo broadcast. The lamp and the emitter-collector of Q_4 are connected in series with the power source. Under normal conditions, and with no stereo signal input, the transistor is biased in an *off* condition. Most of the voltage is developed across the transistor and very little develops across the lamp. A 38-kilohertz carrier signal, when received, is fed from the emitter of Q_3 to the base of Q_4. This signal biases Q_4 so that it conducts, and its internal resistance becomes very low. The result is that the voltage is now developed across the lamp instead of Q_4. The lamp lights, indicating a stereo broadcast. Q_4 acts as a switch in this circuit.

The information just presented about stereo decoders applies to one type of decoding system. There are other types of decoders that are being used in home-entertainment equipment. As technology continues to improve, fewer of these actual circuits are to be seen. A great many of them are being superseded by IC chips containing almost all of the components shown in the schematic in Fig. 16-8. Even so, the knowledge of how a unit works is important to the repair technician. A complete understanding of how a section of a set is supposed to work will make the repair job easier. Merely having to replace an IC or a *module* board goes a long way in the area of rapid repair, but it still is not the entire answer. The person who knows how and why the total set works is more successful in his work.

SUMMARY

Block diagrams are useful to the repair technician. They show the building blocks used in the set. A good idea of signal paths is easily obtained from the block diagram. However, more information than is available from the block diagram is required in order to repair a radio or TV set. This information is to be found on the schematic diagram.

The schematic diagram shows both signal voltages and dc operating voltages. The technician must distinguish between these two voltages when analyzing the set. Most transistor and tube circuits operate as voltage-divider networks. Dc voltage analysis is used in order to locate a malfunctioning part in these circuits. Signal paths are normally shown with the input on the left of the schematic and the output on the right.

Signals handled by an AM radio are received by the antenna and sent to the mixer stage. The local oscillator produces a carrier signal that differs from the broadcast carrier by the IF frequency. The mixer produces a modulated IF carrier which is amplified and fed to the detector. The detector removes the intelligence from the IF carrier and feeds this signal to the audio section of the radio.

The same action occurs in both AM and FM radios. Some radios use an RF amplifier for greater signal strength. Other radios feed the signal from the antenna directly to the mixer stage. Ideally, it is better to have a high signal level fed from the antenna to the set.

FM stereo receivers require additional circuits in order to reproduce the stereo signal. The composite stereo signal is made up of the monaural information $(L + R)$, a pilot signal at 19 kilohertz, and the stereo information $(L - R)$. The stereo information is added to the main carrier as a double-sideband suppressed-carrier AM signal. In order to be detected, the suppressed carrier has to be re-created in the receiver. This is done by using the 19-kilohertz pilot signal and doubling its frequency to 38 kilohertz. Adding this signal to the $L - R$ information permits demodulation.

The FM mono information and the FM stereo information are mixed in the stereo detector section of the decoder. The result of this action is separate right- and left-channel information. This audio information is then fed to right- and left-channel audio amplifiers and to speakers connected to each amplifier.

QUESTIONS

1. Which block of the radio receives the signal from the antenna?

2. How does the local oscillator's frequency differ from the RF section's frequency?

3. How does the local oscillator change frequency in step with changes in the RF frequency?

4. What devices are used as coupling devices in the IF amplifier of a radio?

5. What kind of detector is used in an AM radio?

6. What kind of detector is used in an FM radio?

7. What signals are present on the carrier of a station broadcasting an FM stereo program?

8. How is the stereo information removed from the carrier?

9. What is the purpose of a trap in a radio?

10. Explain how the stereo-indicator circuit works.

Television
Circuit
Analysis

17

In this section of the book, we are discussing circuit analysis. Both audio and radio circuits are explained in the previous two chapters. This chapter tells how the dc operating voltages and the signals are developed and processed in the television set. The black-and-white solid-state set is used as the example in the first part of the chapter. The solid-state set has been selected here because very few black-and-white tube sets are being designed and introduced on the market. The reader should keep in mind that the tube set operates in similar fashion to the solid-state set. One major difference is that higher dc voltages are required for tube sets. Signal paths and other circuitry are essentially the same for solid-state sets.

The latter part of this chapter explains how the color set works. There are great similarities in the signal processing sections of color and black-and-white TV sets. Therefore, these similar sections are not presented again for the color set. That discussion focuses on those blocks that are not found in a black-and-white TV set.

The block diagram for a black-and-white TV is shown in Fig. 17-1. These blocks are the same as those discussed in Chap. 10. This chapter opens up the blocks and looks at the actual circuits found inside them. A complete schematic of this set is shown in Fig. 17-2. Sections of this schematic, related to specific blocks, are reproduced throughout this chapter. The sections relate to the blocks on the block diagram. This approach is used so that the student will more quickly become able to read schematic diagrams.

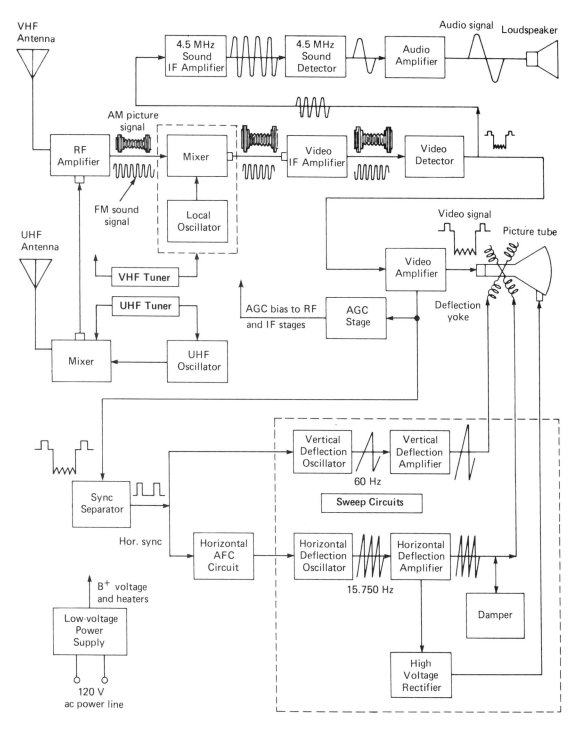

Figure 17-1. Block diagram of the TV receiver, with all signal information waveforms displayed.

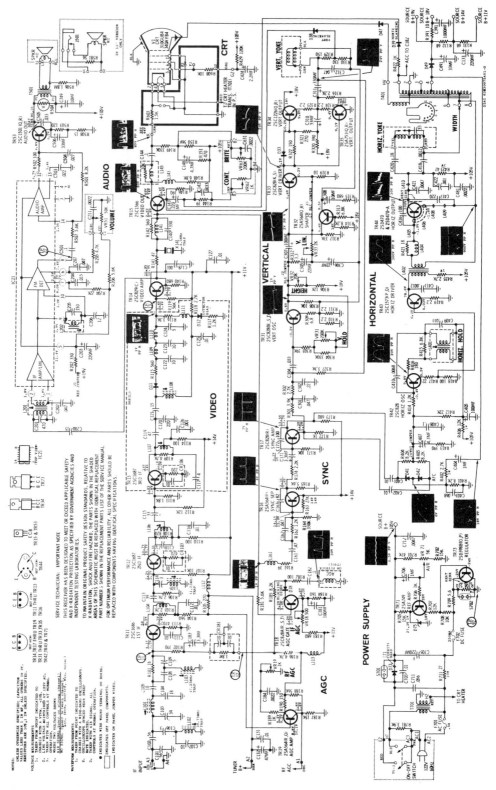

Figure 17-2. Schematic diagram of a solid-state receiver. (Courtesy Quasar Electronics Co.)

The set represented in the figures is actually a Quasar chassis 12 TS-481. This chassis is typical of many solid-state sets on the market. It incorporates integrated circuits, voltage regulation, and has a scan-derived power supply in addition to its regular power supply.

17-1
Power Supply

The main power supply of this receiver is shown in Fig. 17-3. The input to the power supply is 120 volts, ac. The ac source is connected to two circuits. One of these is the primary of T_{701}. When the set is turned off, resistor R_{781} is in series with the primary. This resistor reduces the voltage at the primary of the transformer. One set of contacts in the on-off switch *shorts* this resistor out of the circuit when the set is turned on. The secondary of the transformer is connected to the filament of the CRT. This filament is always *on*. When the set is turned off, the filament voltage is reduced due to the action of R_{781} in the transformer primary.

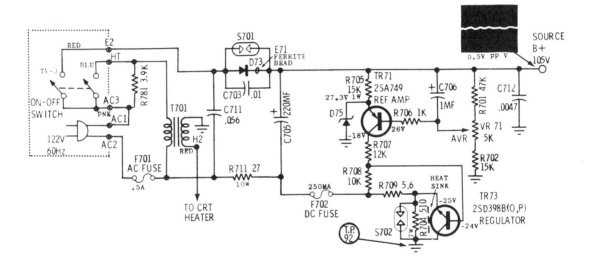

Figure 17-3. The power supply section of the receiver, with regulator circuits shown.

The second connection from the ac power source is through the on-off switch to D_{73}. This diode is used as a halfwave rectifier in order to furnish the +105 volts required for set operation. The term B^+ is used in many electronic circuits to indicate the dc operating voltage. This may be any value. The schematic shows the specific value for the circuit.

The output of the diode is connected to a filter network. This network consists of C_{705} and an electronic regulator circuit. The regulator circuit uses TR_{71} and TR_{73} as active devices to regulate the output voltage. The regulated output supplies the horizontal, audio, and video output circuits in this set.

The regulator circuit works as follows: A voltage-divider network consisting of TR_{71}, R_{707}, R_{708}, R_{709}, and the combination of R_{704} and TR_{73} are connected from common to the positive side of the power source. The purpose of this voltage divider is to maintain 105 volts at the source B^+ terminal. In order to do this, two variable resistances are used. One is transistor TR_{71} and the other is TR_{73}. The regulator transistor TR_{73} is used to control the *reference amplifier* TR_{71}. TR_{73} is used to control the internal resistance of TR_{71}. This transistor is called a reference amplifier because it provides a steady reference voltage to the regulator transistor.

The output B^+ voltage has a second voltage-divider network. It consists of R_{701}, VR_{71}, and R_{702}. Any change in output voltage changes the base bias on TR_{73}. Its conduction changes due to this. If the output voltage starts to increase, TR_{71} conducts less and its collector becomes more negative. The reverse occurs when the output voltage drops. Changes in base bias on the regulator transistor TR_{73} occur when the voltage at the collector of the reference amplifier changes. These changes are used to control the conduction of TR_{73}. Its conduction compensates for voltage changes in the B^+ source voltage, maintaining a constant value of 105 volts.

The other operating voltages for this set are developed in the horizontal sweep section. These scan-derived voltages are further discussed later in this chapter.

17-2
The Tuner

The tuner used in all TV sets is designed to receive all channels. In sets made before the introduction of the varactor tuner, the TV tuner consists of two separate tuning units. One unit receives channels 2 through 13. This is identified as the VHF tuner. The other unit receives channels 14 through 83. This is called the UHF tuner. Both units are mounted together in what is called a *tuner cluster*. The VHF channel-selector switch has a position on it for UHF. A separate tuning knob is used to select the specific UHF channels.

The VHF-UHF tuner is illustrated in Fig. 17-4. It consists of two separate units interconnected by wires from the main part of the set. There are three transistors in the VHF section. One is used as an RF amplifier, one as a mixer, and the third one is the local oscillator. The input from the antenna is connected through a series of inductors to the base of the RF amplifier TR_1. The rotary selector switch shown above the TR_1 is part of the channel selector switch. It has several inductors connected to it. Rotating the switch selects an inductor which, when connected to a capacitor, forms the tuned circuit required for a specific channel. The RF amplifier, local oscillator, and mixer are all tuned in this manner.

The base of the RF amplifier is also connected to the RF AGC circuit in the main chassis. This connection is explained in the section of this chapter discussing AGC circuits. The purpose of this AGC connection is

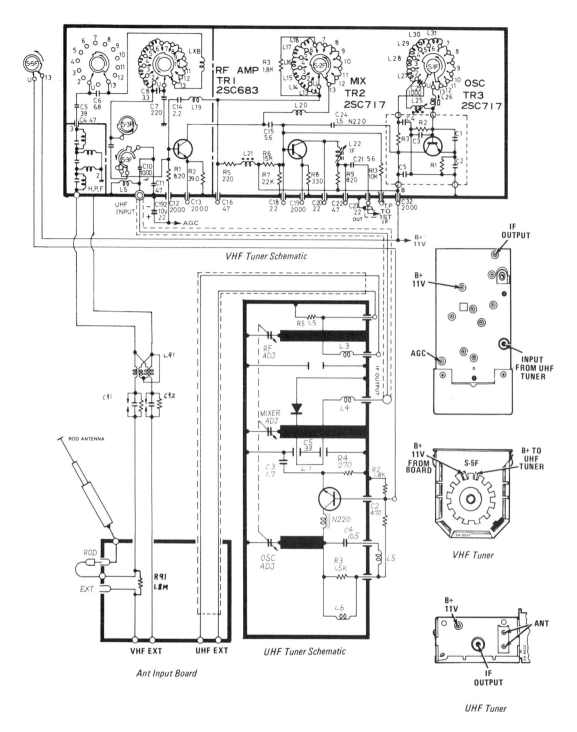

Figure 17-4. UHF-VHF tuner circuits. (Courtesy Quasar Electronics Co.)

to control the amount of amplification in this stage. This is accomplished by changing the bias on the base of this transistor.

The output of the RF amplifier is the collector of TR_1. It is connected through C_{15} and the inductors in the switch shown above the transistor to the base of the mixer transistor TR_2. The output of the mixer is connected from the collector, through L_{22} and C_{21}, to the IF input on the chassis of the set. This connection is made by means of a *shielded* cable.

The local oscillator in the VHF tuner is TR_3. Its frequency is determined by the inductors in the selector switch and the capacitors in the circuit. The output of the local oscillator is connected through C_{24} to the base of the mixer transistor. The output of the mixer is the IF frequency of the set. A close look at the oscillator-inductor selector switch shows an open circuit in the U, or UHF, position. The local oscillator in the VHF tuner is turned off when UHF signals are received.

The UHF tuner does not have an RF amplifier section. It does have a means of tuning the RF signal. The RF signal is selected and fed immediately to a mixer diode. The UHF oscillator output is also connected to the mixer diode. The output of the mixer diode is fed to the UHF input on the VHF tuner. This signal is at the set's IF frequency. With the VHF oscillator disabled, the RF amplifier and mixer in the VHF tuner act as amplifiers for the UHF signal. It is amplified and fed through the VHF tuner to the IF input on the chassis of the set.

Varactor tuning. The varactor is a semiconductor device used as a voltage controlled capacitor. The introduction of the varactor in the TV tuner eliminates the need for switching signals in the tuner. The channel selector switch is used instead to select a dc voltage. The dc voltage is applied to the varactor. The amount of dc voltage is related to the amount of capacitance developed by the varactor. Changing the applied voltage changes the amount of capacitance. The varactor in combination with an inductor forms a tuned circuit.

Figure 17-5 shows a modified block diagram of varactor tuner. The half wheel in the lower portion of the illustration is part of the channel selector switch. This switch has a group of variable resistors connected as voltage dividers. The position of the switch selects a specific voltage. This voltage is fed to the varactor circuits in the RF amplifier, mixer, and local oscillator for the purpose of tuning each of these circuits. The illustration shows a UHF tuner. The same principle works for VHF tuners and combination VHF-UHF tuners. This type of circuit is readily adaptable to tuners using remote controls and digitalized circuitry.

17-3
Video Circuits

Video IF. The video IF section consists of three transistors operating as class A amplifiers. It is shown in Fig. 17-6. The signal is fed to the base of each transistor. Output circuits are connected to the collector of each transistor. The IF signal from the tuner output is connected at the IF input

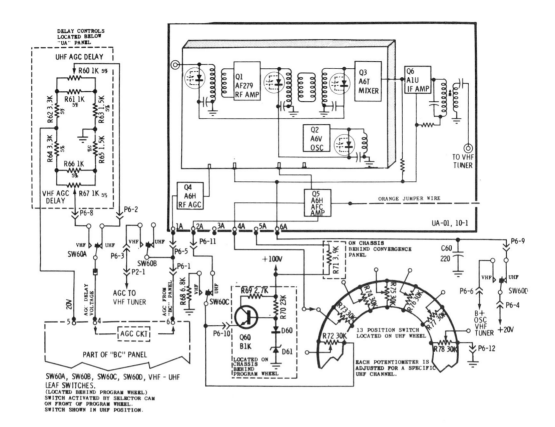

Figure 17-5. The varactor tuner. Voltages developed in the selector switch are applied to varactors to tune the set. (Courtesy Quasar Electronics Co.)

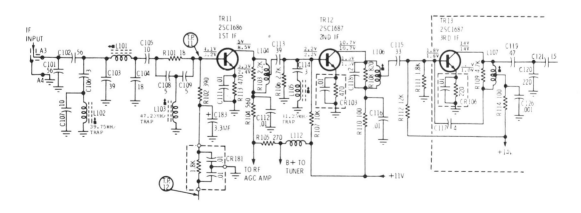

Figure 17-6. The video-IF amplifier section of the receiver.

terminal. The signal is fed through several components to the base of TR_{11}. Operating bias for this transistor is from B^+, through R_{105} and L_{112}, and also from the AGC circuit. The output signal from this transistor is coupled through L_{104} and C_{113} to the base of transistor TR_{12}. Its output signal is coupled through L_{106} and C_{115} to the base of TR_{13}. The output of the third IF is coupled through L_{107}, C_{119}, and C_{121} to the video detector.

Several dc voltage values are shown on the schematic other than those at the elements of the transistors. These values represent connections from power sources or to another circuit. Both the +14-volt and the +11-volt signs indicate a connection to a power source. The label B^+ *to tuner* indicates that the tuner B^+ wire is connected to this point in the circuit. The wire going to the RF AGC amplifier supplies the dc operating voltage from this circuit in the tuner.

Three traps are employed in the IF amplifier. Both the 39.75 megahertz and the 47.25 megahertz traps are resonant circuits used to minimize interference from signals on channels next to the one being received. These traps are used to attenuate, or reduce, signals at these frequencies. The third trap, at 41.25 megahertz, is used to attenuate the sound signal. This is done in order to assure the proper ratio of video to sound in the set.

Video detector.   The video detector is shown to the left of the dashed lines in Fig. 17-7. The detector circuit uses a diode to form a halfwave rectifier circuit. This circuit operates at the IF frequency. The IF carrier is filtered out through the action of C_{122}, C_{123}, C_{124}, L_{109}, and L_{110}. The video signal develops across the detector load resistor R_{116}. The composite video signal is then fed to the video amplifier.

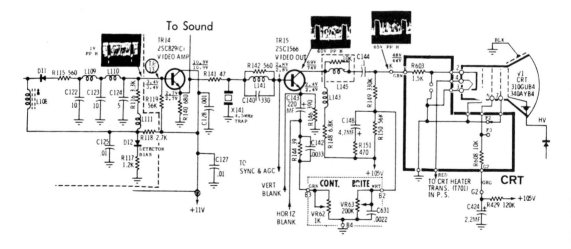

Figure 17.7. The video detector, video amplifier, and CRT wiring of the receiver.

Video amplifier and CRT. These circuits are illustrated in Fig. 17-7, also. They are the circuits found on the right side of the dashed lines. The composite video signal is fed from the detector to the base of the video amplifier, TR_{14}. This circuit has several outputs. One of these outputs is the sound signal. It is coupled from the emitter of TR_{14} to the input of the sound section.

The video portion of the signal passes through X_{141}, a crystal filter used as a 4.5-megahertz sound trap, to the base of the video output transistor TR_{15}. The signal is also fed from the base of TR_{15} to the sync separator and to the AGC circuits.

The video-amplifier collector circuit is coupled through C_{144} to the cathode of the CRT. Signal is transferred to the CRT in this maner. There are two consumer-operated controls on the video output circuit. These are the contrast and the brightness controls. The contrast control is in the emitter circuit of the transistor. It is used to establish a dc bias on the emitter of the transistor. This bias controls the amount of amplification in this stage. The contrast control varies the amount of video information fed to the CRT in the set.

The brightness control is used to establish the dc operating bias at the cathode of the CRT. This, in turn, controls the amount of illumination on the face of the CRT. The brightness control controls the brilliance of the CRT screen.

Two additional inputs to the video output circuit are seen in the emitter circuit of this stage. These are identified as vertical blanking and horizontal blanking. Their purpose is to turn off, or blank, the video output circuit during retrace periods of the scanning signal.

**17-4
Automatic
Gain Control**

IF AGC. AGC, or automatic gain control, controls the amount of amplification in the IF and/or RF stages of the TV set. The amount of AGC voltage is dependent upon the strength of the signal in the IF amplifier. In this set, the AGC system controls the gain in the first IF amplifier in order to maintain a relatively constant signal at the video detector. On strong signals, AGC is applied to the tuner RF amplifier in order to reduce the effects of overloading the circuits. With no, or weak, signals being received, these amplifiers are biased to provide maximum amplification. Increased signal level applies more forward bias to the IF amplifier to reduce its gain. As the signal level increases, the IF gain decreases, until a predetermined level is reached. Beyond this point, AGC is applied to the tuner RF amplifier in order to reduce its gain. This type of AGC is often called delayed AGC because it only controls above a predetermined level.

The circuit in Fig. 17-8 shows key parts of the IF amplifier and the video amplifier as well as the AGC transistor TR_{19}. Consider this circuit under *no-signal* conditions. There are two control circuits shown. One uses TR_{18} to control the base bias on the video IF amplifier TR_{11}. A

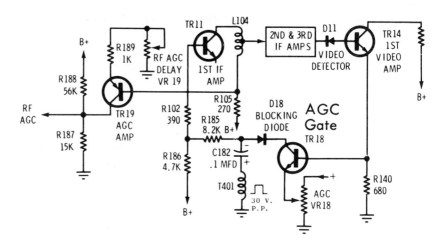

Figure 17-8. The IF and RF AGC-circuits used in the receiver.

positive square-wave-shaped pulse is connected from a winding on the flyback transformer T_{401} through C_{182} to the collector of TR_{18}. This controls the bias on the base of TR_{11}. The RF amplifier is biased by the level of signal from the collector of TR_{19}. This controls the dc voltage at the junction of voltage-divider resistors R_{188} and R_{187}.

With a signal present, a negative-moving signal at the base of TR_{14} reduces the conduction in TR_{18}. The result is more of a positive voltage at the base of the IF amplifier TR_{11}. Conduction through TR_{11} increases and the IF gain is reduced.

RF AGC. The RF AGC voltage is developed at the junction of L_{104} and R_{105}, in Fig. 17-8. These parts are located in the collector circuit of TR_{11}. As the conduction through TR_{11} increases, more of a voltage develops across R_{105}. This large voltage appears as a negative signal at the base of TR_{19}. As a result, TR_{19} conducts and increases the AGC voltage to the tuner.

17-5
Audio System

The audio system in this set is shown in Fig. 17-9. Most of the circuitry is contained in one IC chip. The functions of sound IF, limiting, detection, and audio-signal voltage amplification are performed in the chip. The 4.5-megahertz sound-IF signal is taken from the video detector circuit and coupled to the chip through L_{201}. The chip is shown by the dashed lines and is identified as IC_{21}. Each function is shown by a triangle representing a circuit in the chip.

Amplitude of the audio signal is controlled by the volume control, VR_{51}. The output of the chip is directly coupled to the audio output transistor TR_{53}. This circuit is a single-ended transformer-coupled class-A amplifier. The relatively low impedance of the speaker is matched to the output circuit of the transistor by transformer T_{501}. R_{504} is used

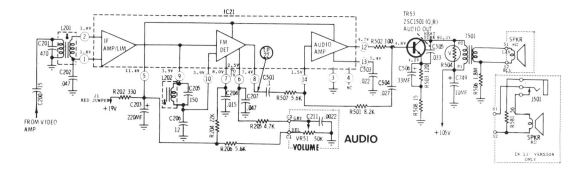

Figure 17-9. The audio system. An IC provides a demodulated and amplified signal for the output stage.

across the primary of T_{501} to protect the output transistor from *voltage spikes.*

17-6
Synchronization

The sync circuit is illustrated in Fig. 17-10. It has two stages. These are the sync separator and the sync amplifier. The waveforms shown on the schematic indicate that only the sync pulses are processed in these circuits. The video information sent from the video detector circuit is not used in these circuits. The purpose of the sync separator is to remove the negative pulses from the composite video signal at the base of TR_{16}. The positive pulses obtained at the collector of TR_{16} are amplified, inverted, and coupled to the base of TR_{17} by C_{171}.

Transistor TR_{17} operates as a phase-splitter circuit. It has two output connections. One provides a positive pulse. It is obtained from the emit-

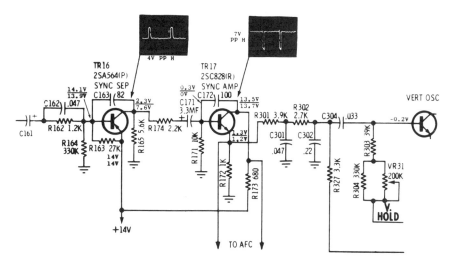

Figure 17-10. Both positive and negative sync pulses are developed in the sync separator.

ter of TR_{17}. A negative pulse is obtained from the collector of TR_{17}. The positive pulse is used to drive the horizontal AFC through C_{402}. It also drives the vertical sweep system through a vertical integrator network consisting of R_{301}, R_{302}, C_{301}, and C_{302}. The negative pulse also drives the horizontal AFC through C_{401}.

17-7
Horizontal Circuits

Horizontal AFC. Automatic frequency controls (AFCs) are used in situations where the frequency of an oscillator must be closely controlled. This is particularly true when the control is keyed to some outside signal. These AFC circuits perform two kinds of work. First, they sense the difference between the oscillator frequency and the desired frequency. Next, they use the voltage developed during sensing to compensate for the off-frequency conditions—the voltage is used to return the oscillator to its proper frequency.

This principle is used in the AFC circuit shown in Fig. 17-11. The horizontal AFC portion of the circuit is to the left of the dashed lines on the schematic. One pulse from the sync separator is fed through C_{401} to the cathode of D_{41}. Another pulse, of opposite polarity, is fed from the sync separator through C_{402} to the anode of diode D_{42}. The diodes are connected in series. A third pulse, taken from the horizontal output (flyback) transformer, is connected to the junction of the two diodes. This pulse is fed through a wave-shaping network in order to produce a sawtooth waveform.

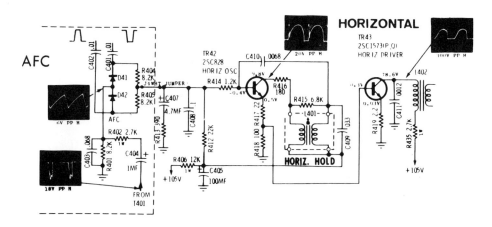

Figure 17-11. Horizontal frequency control is accomplished in the AFC section. Generation and amplification of the horizontal signal produce a 300-volt peak-to-peak signal.

When the oscillator frequency is correct, there is zero voltage produced at the output of the circuit. If the oscillator operates above the desired frequency, a positive voltage is produced. A low frequency oscillator produces a negative voltage at the output. This works in a manner very similar to that of the discriminator circuit used for FM detection.

The output of the AFC circuit is fed to the base of the horizontal oscillator transistor TR_{42}. The oscillator feedback circuit is made up of C_{410} and the secondary of L_{401}. The oscillator output is fed from the emitter of TR_{42} to the base of TR_{43}, the horizontal driver transistor. The signal amplitude is increased from 20 volts peak-to-peak to 300 volts peak-to-peak in this stage, or 15 times the size of the input signal. The output of the horizontal driver is coupled through L_{402} to the horizontal output stage.

Horizontal output. This stage serves several purposes. The outputs from the horizontal output stage include the high voltage required for the CRT, the horizontal AFC pulse, the blanking pulses, the sweep for the deflection yoke and, in some sets, the dc operating voltages for parts of the receiver. The schematic for this stage is shown in Fig. 17-12. Most horizontal output stages operate as class C amplifiers. The stage is normally biased to operate at a point below cutoff. This means that the output transistor is not conducting. Pulses of signal voltage develop in the horizontal oscillator and driver stages. These pulses are used to bias the horizontal output transistor, TR_{44}, into conduction mode. The circuit involved starts at common, goes through the flyback transformer T_{401}, through the emitter-collector of the transistor, through R_{423}, and connects to the B^+ power source. A magnetic field develops in T_{401} when the transistor conducts.

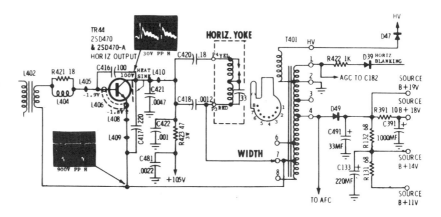

Figure 17-12. The horizontal output stage drives the yoke, and provides both high and low operating voltages for other circuits.

A negative pulse from the horizontal driver and L_{402} cuts off the output transistor. The magnetic field built up in the flyback transformer T_{401} collapses. The collapsing magnetic field develops a voltage in the high-voltage winding of T_{401}. This voltage is rectified by diode D_{47} and sent to the CRT.

A low voltage dc source is obtained from the flyback transformer in this set. This *scan derived* voltage is taken from a tap off the primary winding of the transformer. The connection point is number 4 on the schematic. The frequency of the voltage is the same as the horizontal frequency of the set, that is, 15,750 kilohertz. These high frequency pulses are rectified by diode D_{49} and filtered by C_{491}. This provides about 19 volts dc. The 19-volt source is distributed in order to obtain sources of 11, 14, and 18 volts dc. These sources are used to provide operating voltages for other circuits in the set.

The other major circuit operated by the horizontal output circuit is the horizontal deflection-yoke circuit. The purpose of the deflection yoke is to provide a changing magnetic field which will force the electron beam in the CRT to move across the face of the tube. The pulses of signal developed in the horizontal output stage during conduction develop into a varying magnetic field in the horizontal-yoke winding. This magnetic field forces the CRT electron beam to *scan* the face of the tube horizontally.

The yoke is positioned around the neck of the CRT. It has two sets of windings. One set controls the horizontal scanning. The other set controls the magnetic fields for vertical scanning. These two sets of coils interact to make the electron beam move to every position on the CRT.

17-8
Vertical Circuit

The purpose of the vertical circuit is to develop the necessary waveforms of sufficient amplitude to operate the vertical windings of the deflection yoke. The complete vertical circuit is shown in Fig. 17-13. The oscillator transistor TR_{31} develops a sawtooth wave. The frequency of this wave is 60 hertz. Its purpose is to develop a current through the yoke.

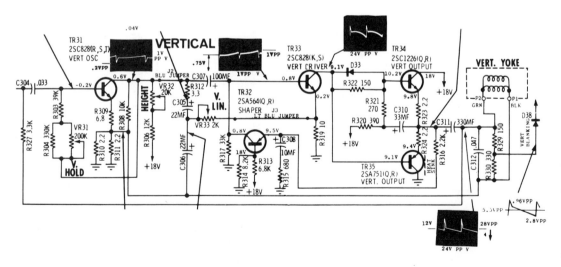

Figure 17-13. Vertical signals are developed in this section in order to drive the yoke.

The sawtooth waveform modified in a wave-shaping network is shown near TR_{32} in the illustration. This wave is amplified by TR_{33} to about 24 times its size at the base of this transistor. The output of TR_{33} is directly coupled to the complementary-symmetry output circuit using transistors TR_{34} and TR_{35}. This power output stage is used to develop the required current for yoke operation. The emitters of the output transistors are used for the output path of the signal. The modified sawtooth wave is coupled to the deflection-yoke vertical windings through C_{311}.

Three controls are found in this section. The *vertical-hold* control is used to adjust circuit values in order to lock the vertical oscillator frequency with the vertical synchronizing pulse. This keeps the picture from rolling. The *vertical-linearity* control adjusts the spacing of each horizontal line. When these spaces are equal, the vertical sweep produces a picture that faithfully reproduces the original image seen by the TV camera. A nonlinear picture would result in stretched portions of the picture. The third control in this section is the *height* control. This control adjusts the amount of amplification in the vertical circuit. Too much amplification produces overscanning and, as a result, only the center part of the picture is seen on the screen. Too little amplification produces a picture which does not fill up the entire screen of the set.

The reason that the waveform produced is nonlinear is that scanning speed changes as the beam moves across the face of the CRT. The distance from the electron gun on the CRT to the face of the tube is greater at the edges than it is in the center. Therefore, the scanning current is changed in order to maintain a constant scanning rate at the face of the tube.

This completes our analysis of black-and-white TV receiver circuits. Keep in mind that different manufacturers may use different circuitry in order to reproduce the TV picture. Even so, the circuits we have discussed are the basic building blocks for almost all TV receivers. Signals move through the various sets in the same manner. The outputs are the same for all sets. Operating voltages differ among sets. Signal amplitudes also vary among sets. In all sets, the circuits used are either amplifier, oscillator, or rectifier circuits. Operation of these basic circuits produces the desired outputs from the receiver.

17-9
Color Circuits

The compatible color-TV receiver has many blocks that operate in the same manner as those of the black-and-white receiver. Figure 17-14 shows the block diagram of a color receiver. The blocks outlined by dark lines are added to the basic black-and-white set along with the three-color CRT.

Color signals are usually divided into two groups. One group consists of those signals relating to color synchronization. The other group is called the color information processor. The purposes of the color sync system are to reconstruct the color carrier in the receiver and to synchronize the color oscillator in the receiver with that of the transmitter.

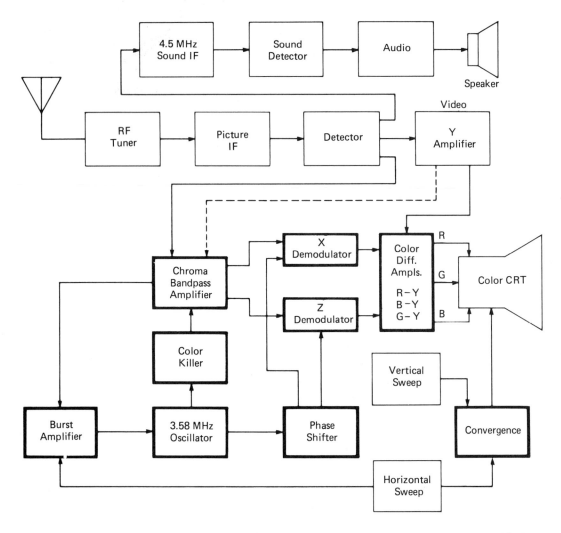

Figure 17-14. The color blocks of the receiver are outlined by heavy lines.

The color carrier is required in order to demodulate the color information. This demodulation process is similar to that used in the FM stereo decoder.

Color processing blocks used in a Quasar TS-942 chassis are illustrated in Fig. 17-15. The panel shown is one of several used in this chassis. Follow the signal through this panel. The composite video signal, including color information, is fed to the circuit from the video detector. It is connected at terminal 2. A *delay line* (D.L.), is used to hold back the signal for a few microseconds. This is done in order to process the color information correctly. The video and color information have to arrive at the color demodulator at the same moment in order to reconstruct the color pic-

278

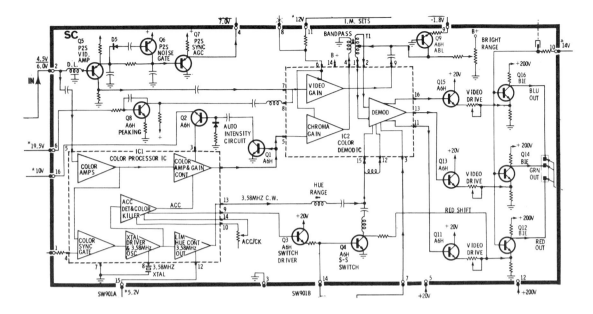

Figure 17-15. Color processing circuitry. (Courtesy Quasar Electronics Co.)

ture. The video signal is fed to Q_5, the video amplifier transistor. From this transistor it is fed to a video gain block (amplifier) in the color-demodulator chip, IC_2. The signal is coupled to the demodulator block by means of bandpass transformer T_1. Let's leave this path now, and return to the input terminal.

The color signal is coupled to the color processor TC_1 and fed to the color-amplifier gain block. Two outputs are connected to this block. One feeds the color amplifier and gain control block. From this block it is connected to a chroma gain block in the color demodulator chip IC_2. This signal is coupled through T_1 to the demodulator. The other output from the color amplifier block is the color sync signal. It is connected to the color sync gate. The output of this block drives the color oscillator block.

In this set, the output of the 3.58-megahertz color-oscillator block controls the color-killer circuits, the automatic color control, and the hue. The output of the hue block feeds the demodulator. The output from the color killer feeds the color-amplifier gain block. Under a no-color-signal condition, the output of the color killer block is used to turn off the color processing blocks in the receiver.

The demodulator uses the video information and the 3.58-megahertz carrier from the oscillator in order to reconstruct an amplitude modulated signal. After the signal is reconstructed it is demodulated into three signals. These signals represent the three basic colors of red, blue, and green. The three outputs are directly coupled to the bases of their respective color-video amplifiers, Q_{11}, Q_{13}, and Q_{15}. The signals are connected from the emitters of these transistors to the emitters of the color amplifier transistors Q_{12}, Q_{14}, and Q_{16}. The bases of these transistors are parallel wired to the set's brightness control. Each transistor output is connected to the collector of the transistor. From each collector, the signal is directly coupled to the correct cathode of the color CRT.

These circuits may be constructed using discrete components. The state of the art in electronics is such that much of the circuit may be provided as one integrated circuit. Most color set manufacturers now use the IC to replace the several transistors required to make an equivalent circuit. Many of these same manufacturers produce modules on boards which contain a group of related circuits. Initial service for these sets requires the identification of the specific board on which the malfunction is located. Changing the board repairs the set. The board may then be returned to the manufacturer for repair and/or remanufacturing. Under the proper conditions, it is possible for the technician to repair a board in the shop.

SUMMARY   The composite video signal is processed through the TV receiver in a manner similar to signal processing in other receivers. A solid-state black-and-white TV receiver is used as the model in this chapter. This set is typical of those produced in recent years.

The power supply in the set uses a halfwave rectifier circuit for the dc operating voltages. This power supply feeds the horizontal, audio, and video sections of the set. Voltage control is developed by use of an electronic voltage-regulator circuit. This circuit senses changes in output voltage. These changes control the internal resistance of a voltage-regulating transistor. Changes in the internal resistance of the regulator transistor cause the dc output voltage to change in order to maintain a specific voltage level.

Ac voltage is provided for the filament of the CRT. This is accomplished by use of a filament step-down transformer in the power supply. This set uses a resistor to reduce filament voltage on the CRT when the set is off.

The tuner is a typical, combination VHF-UHF unit. It selects a series of inductors by means of a switch. The inductors are connected with a capacitor by the switch. The combination inductor-capacitor circuit tunes the desired channel. The VHF tuner has three blocks. These are the RF amplifier, local oscillator, and mixer. These blocks function in a manner similar to those in the tuner in a radio in order to produce a modulated IF carrier for the set.

UHF tuners use a mixer diode and a local oscillator to provide a modulated IF carrier-signal. This signal is fed to the RF amplifier of the VHF tuner. The IF signal is amplified by the VHF tuner, RF amplifier and the mixer. The VHF local oscillator is turned off during UHF reception. When this happens the mixer stage acts as an amplifier. As a result, the modulated IF signal from the UHF tuner is amplified in the VHF tuner and sent to the receiver's IF amplifier strip.

Recent improvements in turners include the use of varactors for tuning purposes. The varactor removes any RF from the selector switch contacts. The result is a better tuning system which requires a lot less maintenance. The varactor's capacitance is changed by varying a voltage placed across the device. Using the varactor in combination with an inductor produces a voltage-controlled tuned circuit.

The video IF amplifier in the receiver provides large amounts of gain for the signal. The signal passes through a series of common-emitter amplifiers to the video detector.

The video detector employs a halfwave rectifier circuit to remove the video information from the modulated IF carrier. This stage is the take-off point for the sound signal, the synchronizing pulses, and the AGC signal.

The output of the video detector feeds the video amplifier circuits. Signals in this set are amplified in a common-collector video-amplifier circuit and sent to a common-emitter video-output circuit. The signal is capacitive coupled to the cathode of the CRT. Contrast and brightness controls are located in the video output circuit.

IF AGC is developed from signals sent from the video detector. The AGC circuit combines a video signal and a pulse from the horizontal output transformer to produce a control voltage for the first IF-amplifier stage. After a predetermined signal level is attained, the AGC voltage is sent to the RF amplifier in the tuner. This is done to provide additional amplification for weak signals. The purpose of AGC is to maintain signal level at the video detector.

The audio system in this set is found on an IC chip. Sound-IF amplification, detection, and voltage amplification occur in the audio chip. The output from the chip is fed to a common-emitter power-amplifier circuit. From the power amplifier, the signal is coupled through an audio output transformer to the speaker.

The sync circuit develops a pulse. This pulse is used to synchronize the horizontal and the vertical oscillators with those at the TV station transmitter. A phase-splitter circuit is used to develop both a positive-moving pulse and a negative-moving pulse.

Horizontal AFC is a frequency control circuit. It employs both the positive and negative pulses from the sync circuit and a pulse from the horizontal output transformer. These three pulses develop a control voltage for the horizontal oscillator. Frequency changes produce a shift from the desired control voltage. This changed voltage is fed to the horizontal oscillator to correct for the frequency shift.

The horizontal oscillator produces a modified sawtooth wave signal. This signal is used to drive the horizontal output stage. The circuit uses two common-emitter amplifiers in order to increase the signal from 20 volts to 300 volts.

Horizontal output stages are used to develop high voltage for the

CRT, drive the horizontal yoke windings, and provide pulses for control circuits. This set develops a *scan-derived* voltage from the horizontal output transformer in addition to that output's other functions. The horizontal output stage operates as a class C amplifier. It is biased well below the point of cutoff. A magnetic field develops in the primary winding of the output transformer during conduction. The field collapses during cutoff. When this occurs, the collapsing field generates a high voltage in the secondary of the transformer. This voltage is rectified and used at the CRT.

The vertical circuits develop a sawtooth wave in the oscillator section. The signal is amplified in the driver transistor. It is then fed to a complementary-symmetry class-B output stage. The output signal is capacitive coupled to the deflection yoke.

The color set uses circuits that are very similar to those found in the black-and-white set for everything but color signal processing. Color information is classified as either color sync or color picture information. Sync signals control the frequency of the 3.85-megahertz color oscillator. This oscillator signal is used to provide a carrier for signal detection and to operate the color killer section. Color picture signals are amplified in a bandpass amplifier and sent to the demodulators. Video and color information are combined in the output of the demodulators. The result is a red, a blue, and green signal, each containing video information. These three signals are fed to common collector amplifiers and from these to common emitter amplifiers. The outputs of the three color-amplifiers feed the color signals to the cathodes in the color CRT. The end result is a color picture.

QUESTIONS

1. Why is a regulator circuit used in the power supply circuit?

2. What is the purpose of the VR_{71} control in the power supply shown in Fig. 17-3?

3. What two signals are present at the input to the mixer stage in the tuner?

4. Why is the local oscillator in the VHF tuner disabled when receiving a UHF signal?

5. What type of amplifier circuits are used in the video IF stages?

6. Describe AGC action in both the IF and RF stages of the set.

7. How much signal gain is there in the video amplifier section?

8. What is the purpose of AFC in this set?

9. Could AFC circuits be useful in other kinds of receivers? How and where would they be used?

10. What advantages are there when a scan-derived power source is used?

11. What is the function of the vertical hold control?

12. Why is a delay line used in a color set and not in a black-and-white set?

13. What is the wave shape of the signal obtained from the color oscillator?

14. What is a *gain block*?

15. How does the color-killer circuit work?

The Troubleshooting Approach

18

There comes a moment in the career of the novice repair-technician which, for him, can be called "the moment of truth." This is the time when everything has to come together. The technician is assigned a set to repair. He has studied all the material in this book. Now, the technician is about to apply book knowledge in order to repair the set. At this point, the crucial thing for him is knowing where to start and what logical steps should be taken in order to make a successful and rapid repair.

There is no easy formula for repair of all electronic devices. Most units are made from literally hundreds of individual components. Any one or more of these may fail, causing malfunction of the set. Finding the specific component is not always a simple job. One certainly doesn't want to take several hours fishing all around in the set for the bad part. Speed is related to income in most repair shops. The successful technician wants to learn how to make a rapid and correct diagnosis of the fault. Then, based on knowledge of how the circuit works, the set is quickly repaired.

One of the most important tools the technician has is his mind. The mental attitude of the person doing the repair is very important to the success of the repair. Some people have facetiously accused certain sets of being human. The technician may sometimes imagine that the set is an animate device with a mind of its own. It isn't. The set is made up of many individual parts. These parts do *not* have the ability to think. The technician does have this ability. The set was working when it was made. The engineers designed a working apparatus. The set was demonstrated to the customer in the store, and it worked there, too. Now, some part has failed. By use of a logical, systematic approach the technician can correct the malfunction and return the set to its original working condition.

The technician does have to think. He has to be able to see what it is that's not working right. Even his sense of smell should be good enough for him to be able to detect overheated parts. He uses a block diagram and the knowledge of how the set is supposed to work in order to locate a specific area of trouble. Once this area is located, he uses the schematic diagram and test equipment to locate the specific part which is bad. Once this part is replaced, the set is tested again in order to be sure that the repair was correct. Then, the set is ready to be returned to the owner.

18-1
Logic of
Troubleshooting

The primary concern of any technician is to know the system requiring repair. Without this knowledge, it is impossible to determine whether the system is working correctly. There are a few basic systems used in consumer electronics. These systems include amplifiers, and AM radios. The block diagram for an AM radio differs slightly from that of an FM radio. The block diagram for a TV set has many blocks that operate exactly in the same manner as those found in an AM, or FM, radio. Amplifiers are used for audio, video, and IF signals. Each operates in a similar manner. The main difference among these amplifiers is in the frequencies handled by each of them. It would be very wise for the technician to memorize the basic block diagrams for an amplifier, a radio, and a TV set. These block diagrams can be used for most sets to aid in diagnosis of problems. Knowing these block diagrams greatly aids in development of a logical, systematic approach to troubleshooting.

As the technician gains experience, he will find that there will be some sets with *different* circuits. However, these circuits will become familiar to him as the sets are worked on and repaired. Knowledge of the basic systems, including signal paths, will help make all sets easier to repair.

Building blocks. Each and every electronic system is made up of subsystems, or *functions*. These functions have to operate properly if the total system is to work. Another name for these functions is building blocks, or blocks, and these have been discussed in this book in great detail. The blocks, put together form a system diagram. Knowledge of how the blocks form the system, coupled with knowledge of how the signals flow through the blocks, is fundamental to the process called *system analysis*.

Diagnosis of a system malfunction is very simple. The first step is to know the blocks. Operation of the set will help to determine which of the blocks are working correctly. These blocks are then eliminated from consideration. The remaining block or blocks are then assumed to be not working. It is in these blocks where the fault is to be located. Following the identification of a specific malfunctioning block or blocks, the technician turns to the schematic diagram and looks at the circuitry which makes up the block. The stages involved in the block are analyzed in order to isolate a specific nonoperating stage. The stage itself is further broken down into specific circuits, which may be labeled as input circuits or output circuits within the stage. Finally, the circuit is analyzed in order to locate a specific component which has failed. Development of this kind of a methodical approach goes a long way in preparing a person for success in electronic repair work. The pattern used to isolate a bad component is shown in Fig. 18-1. This illustrates the concept of moving from the set to the blocks, then to the stages, and finally to the part in order to repair an electronic device.

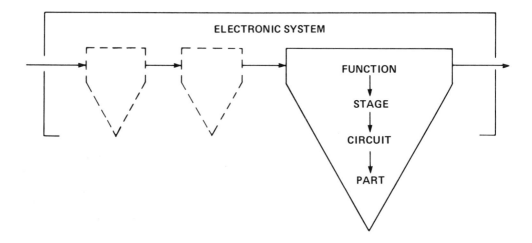

Figure 18-1. Pattern of thinking about a repair problem should proceed from the large system down to the small part.

Technical information. Few technicians ever memorize the electronic circuit diagram for an entire set. This type of knowledge is too much to expect to commit to memory. There are too many different set circuits in use for a technician to be able to depend upon his memory for the exact circuit. Certain basic circuit configurations should be learned. These include rectifier circuits, filter circuits, and amplifier circuits. Knowledge of how each basic circuit, described in Chap. 14, works is necessary for the successful technician.

Using technical information is necessary for successful repair. Finding this information sometimes proves more of a challenge than the actual repair of the set. There are several sources of technical information from which the technician may draw. One of these sources is the set's manufacturer. Most manufacturers of common electronic products have established national service-parts distribution systems to aid the technician in obtaining both technical literature and proper replacement parts for the set. Manufacturers who do not have local distributors in major cities usually have an address from which literature may be ordered.

There are other sources of technical literature as well. These are booklets published by private companies and sold through local, independent, electronic parts distributors. This type of literature covers a great number of set manufacturers. A visit to the local electronic-parts distributorship will reveal the information available and the extensiveness of this information.

18-2
Testing Equipment

Three broad categories of electronic test equipment are used in the repair of the electronic device. These are: *signal generating, signal tracing,* and *voltage measuring.* These are not the only types of equipment available. There are other types available for specific situations, which are very useful to the technician. These three groups are discussed in order to show general procedures for troubleshooting. The general procedures are used to help isolate a defective stage in the set.

Specific components are often tested with such devices as transistor or tube testers, capacitor checkers, and CRT testers. These units provide information about the specific part. They may be used *after* the area of trouble is isolated in the set.

Signal tracers. Use of the signal tracer presumes that a good signal is available at the input to the set. Somewhere in the set the signal is blocked due to a malfunctioning stage. There are two signal tracing devices in general use. One is called, simply, a *signal tracer.* It is an electronic amplifier and includes a speaker as its output transducer. Starting at the input of the set in question the signal is traced and reproduced on the tester's *speaker.* Connections are made to the set being tested by means of test probes. The signal tracing method moves from the input to the output of the set. The trouble area is found when the signal is not heard in the tracer. Normal procedure is to move from the input of one stage to the input of the next stage. If the signal is good at one input and bad at the next input, then the area of trouble has been located. Other test equipment is employed at this time in order to locate the component which has failed.

This same method is often used with an *oscilloscope,* which is the other signal tracing device in general use. The oscilloscope displays the electronic signal in visual form on the face of a CRT. The manufacturer's literature is used as a reference. Waveforms shown on the literature are compared with waveforms on the scope. When there is a major difference, the area of trouble has been found.

This procedure is illustrated in Fig. 18-2. Waveforms are obtained at specific test points in the set. This is where knowledge of the signal paths and a general idea of the kind of waveforms normally found in the set are

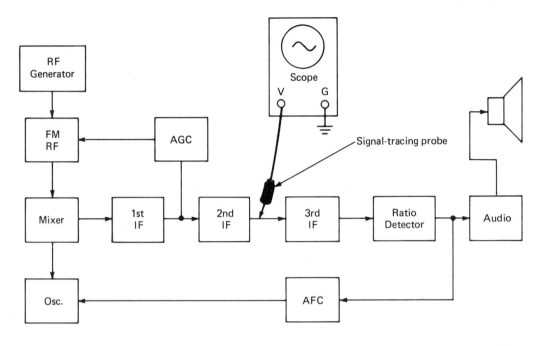

Figure 18-2. Use of a signal tracer. The scope probe is moved from the RF section towards the audio, until the signal is lost.

important. Using the schematic diagram, test points are identified. The scope probe is placed at these test points, in turn. A comparison between the actual waveform obtained and the expected waveform is made. If the two compare, then the probe is moved to the next test point. Testing is done starting at the input and working toward the output of the set until the trouble is located.

These procedures are slow and often tedious. There are ways of saving steps, and these are discussed later in this chapter. It is important to grasp the concept of moving from the good input toward the output until a bad stage is located. This procedure holds for any signal tracing method or equipment used.

Signal generators. A signal generator is an electronic test instrument that is able to re-create a signal originally found in the set. There are *audio* signal generators, *RF* signal generators (with and without modulation capabilities), *FM stereo* signal generators, and *TV* signal generators. These units range from a simple, one-frequency device to units capable of reproducing any one part, or all, of the complex color-TV signal.

The procedure, when using a signal generator, is to start at the output of the set and work backward until the signal is *lost.* An example of this procedure would be in the location of a bad stage in the video amplifier of a TV set. The output device is the CRT. The input to this block is from the video detector diode. A signal generator capable of producing the video signals is used. The test probe is connected to the CRT input element and the set is turned on. If the CRT is good, the video information from the generator will be seen on the CRT screen. Therefore, the probe is then moved back toward the front of the set (that is, to the video detector), and from one stage to another. Each good stage will produce an image on the CRT. When there is no display with the generator connected, the area of trouble has been located. This procedure works in any set. The key is to use the proper type of generator in order to obtain the desired output from the set.

Not all generators produce visual or audio signals. Some generators will reproduce the *pulses* found in a TV set. These could be sync, vertical, or horizontal pulses. Substitution in the set for these signals also helps locate specific areas of trouble.

Voltage testers. A voltmeter is a very helpful device in circuit analysis work. This meter is used to measure the amplitude of the voltage at a point in the set. Both dc and ac voltages may be measured. Here, again, the technical literature is used as a reference. Measured voltages are compared to the values on the schematic. If there is a major difference then a problem area is found. This type of test is not normally one of the first tests made on a set unless one is checking the power supply block. Waveform testing cannot be done with a voltmeter, as this device cannot display more than one value at a time.

Circuit analysis is done after voltage tests are made. The rules that apply for series and parallel circuits as well as open and short circuits are used in order to locate a nonworking component. These rules apply for both operating voltages and signals. Troubleshooting is very easy when the rules are used in order to locate a bad block, stage, or component in the set.

18-3
The Rules for
Troubleshooting

The purpose in a logical troubleshooting procedure is to identify a specific component which has failed. The approach used is to start with a large unit, such as the complete set, and reduce the area of trouble by excluding the blocks which work properly. The technician knows how the set is supposed to work. Elimination of the working blocks helps him focus on the nonworking blocks in which the trouble is to be found. It is helpful to place *brackets* around the area of trouble in the set on the schematic, or mentally. Making a specific test tells the technician whether a block is working. The test results are then used to move the bracket to a spot which reduces the size of the trouble area. The brackets are moved after each test until only a small area remains. The trouble is in this small area.

Knowledge of what to bracket, where to test, and how to move the brackets helps expedite the troubleshooting process. There are five circuit configurations used for both signal paths and dc circuits. The circuit configuration may be linear, divergent, convergent, feedback, or switching. Specific rules apply for each of these. The technician determines the kind of system used by the information found in the technical literature. The rules for that circuit are then used in order to reduce the trouble area and locate the defective part. Each circuit troubleshooting rule assumes that there is a good input to the circuit and a bad output from the circuit. The rules are used in order to take some major steps to rule out working parts of the set and focus on the nonworking parts.

Linear paths.   Linear circuits are series circuits. This is illustrated in Fig. 18-3. The troubleshooting rule for these circuits is often called the *half-split* rule. The procedure for using this rule is to make a measurement at or near the middle of the linear circuit. If the signal is good at this point then the trouble is between the output and the test point. This method eliminates from suspicion half of the circuit with one test. In a set having 10 or more stages connected in the linear path configuration, it could be very time consuming to measure each stage input. This method reduces the time involved in locating a bad stage or block.

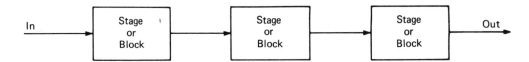

Figure 18-3. A linear path system. Signal flow is indicated by the arrows.

After the first test one of the two real or imaginary brackets surrounding the complete linear circuit on the schematic is moved to the test point. A second test is made on the balance of the circuit. This test uses the same rule. The test is made at, or near, the midpoint of the remainder of circuit. The results are interpreted in the same manner as the first set of results. The bracket is moved again. Each time this is done the brackets come closer to each other until only one stage is left. This stage is defective. At this point, additional testing equipment is used to locate a bad component.

Divergent paths.   The divergent-path circuit is shown in Fig. 18-4. This is a circuit with a common feed point and multiple outputs. The rules for this type of circuit state that, if the output of one block is normal, then the stage or block feeding this output must also be normal. If this is true then the problem has to be in one of the other branch circuits.

If, on the other hand, the test made at the common feed point shows that the signal is not normal, then the problem is in the block or stage

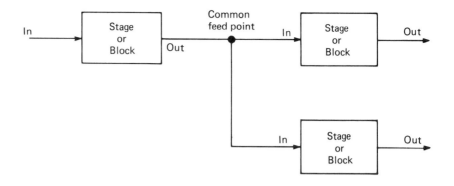

Figure 18-4. Divergent-path system. The signal is fed from a common point to two or more blocks.

before the point of divergence. This type of test quickly reduces the trouble area.

Convergent paths. This circuit has two or more inputs and one output. It is illustrated in Fig. 18-5. The mixer-converter stage of a radio or TV set uses this circuit path. The rules for this circuit state that, if the desired output is normal, then the entire circuit is working. If the desired output is not normal, certain tests may be made in order to locate the defective stage or block.

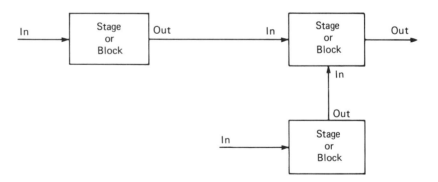

Figure 18-5. Convergent path system. Two or more signals meet at a common point or block.

Each input path is tested at the point where the signals enter the common stage. If both inputs are good, the trouble is in the common stage. If one of the inputs is not good, then the trouble is in the stage before that input.

291

Feedback paths. A feedback circuit is shown in Fig. 18-6. In this circuit, some of the output is fed back to the input for purposes of control. Typical feedback paths in consumer electronic devices include AGC circuits and those found in some oscillators. The rules for testing this circuit state that the feedback path should be broken (or opened). If the output signal returns to normal then the trouble is in the feedback circuit. If there is no difference when the feedback circuit is disconnected, the trouble is in the forward path.

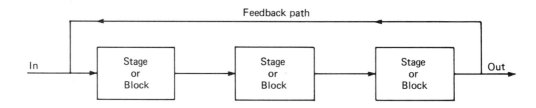

Figure 18-6. Feedback path system. A portion of the output signal is returned to the input for purposes of control.

Switching circuits. This type of circuit may have multiple inputs and a single output, or, in other instances, it may have a single input and multiple outputs. Let's look at the multiple input circuit in Fig. 18-7. This circuit could be that of a high fidelity system. The inputs could then represent an AM tuner, an FM tuner, or a tape deck. The output block would represent the amplifier. The switch is used to select a specific input.

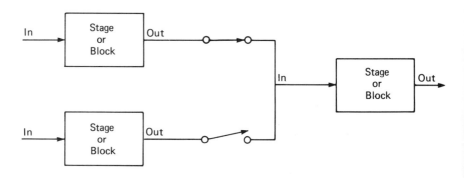

Figure 18-7. Switching path system. One input is selected from two or more sources by use of a switch.

The rules for this circuit state that, if the output is not normal, another input should be selected. If the output still is not normal then the trouble area is in the common stage after the switch. If switching to a second input clears up the trouble, the problem is located in the first input block.

The circuit rules for a single-input multiple-output switching-circuit are similar. If one output is abnormal, switch to another output. If the second output is also abnormal then the trouble area is before the switch in the common block. If the trouble is not present when the second output is connected, the trouble is between the switch and the output of the first block.

Use of the rules for these five circuits helps to simplify the troubleshooting procedure. The next rule, which applies to each of the rules, is to use the information obtained when making these tests. Logical troubleshooting includes analysis of the information obtained during the troubleshooting processes. Use of a logical, systematic troubleshooting procedure makes electronic repair a challenging and rewarding experience.

It is very easy to want to skip steps in the troubleshooting process. However, skipping steps can totally defeat the technician's purpose. The missed step may be the one which will focus on the specific part or block that is not working. A definite procedure is used in order not to miss steps. This procedure is called a *troubleshooting funnel* because it goes from a broad, general area to a specific part. This funnel approach is illustrated in Fig. 18-8. The dashed lines are there to indicate that a set with more than one problem should be approached by solving one trouble and then returning to the beginning to solve the second trouble.

Troubleshooting also proceeds according to a priority of probability. This approach, as illustrated in Fig. 18-9, helps the technician play the odds on failures. This chart highlights failures in order of likelihood occurrence. There have been more active device failures than failures of passive devices. This is not to say that passive devices do not fail. It says that the odds on the active device's failing are greater. Because of this, some devices should be tested before others in the set.

A review of the steps involved in logical troubleshooting identifies four major steps. These are:

1. Obtain information.
2. Interpret information.
3. Narrow limits of fault area.
4. Isolate to identify actual fault.

To begin with, there must be symptom recognition. This is obtained through observation. Look at the set to see if any wires are broken. Are any parts scorched? Is any part broken? Does it light up? Do the tube filaments light? Is sound present? What happens when the controls are turned? Do they do their normal job? These are the kinds of details that help the technician in the diagnosis of problems. This information is added to the problem's description given by the customer.

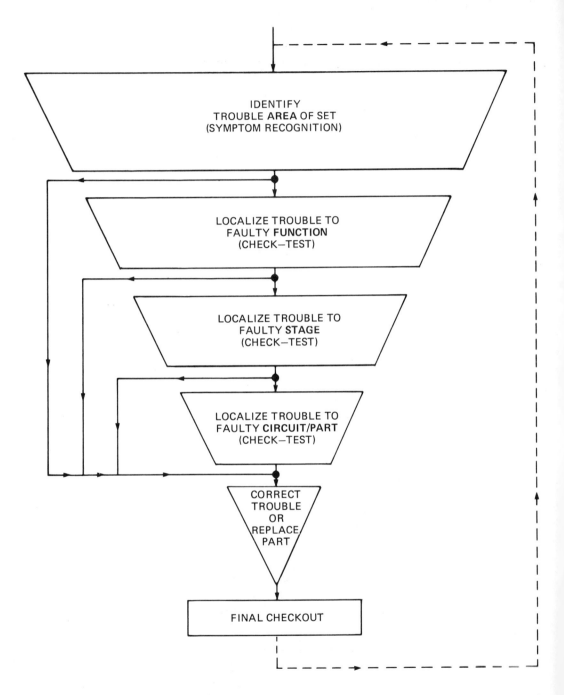

Figure 18-8. An information-seeking funnel as it relates to logical troubleshooting.

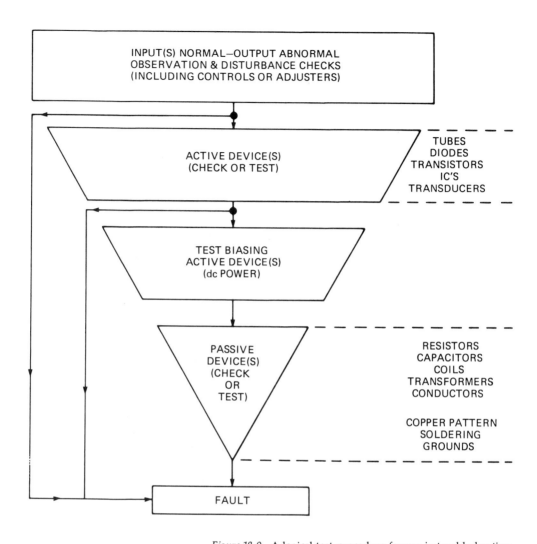

Figure 18-9. A logical test-procedure for use in troubleshooting.

Analysis of this data speeds the repair. Once the data is interpreted, blocks may be eliminated from concern. The remaining blocks form the area of suspicion for the problem. Test equipment and technical literature are used at this point in order to locate the component that is malfunctioning.

18-4
Parts Replacement

Once the defective component is located, it has to be replaced. The first step in the replacement process is the identification of the part needed. Some parts, such as resistors or capacitors, employ a standard system for marking their values. In general, these are replaced by parts with the same values. The technical literature is used in order to identify actual values such as voltage or wattage ratings.

Other parts are rather special in design or shape. They are specifically designed to work in a particular kind of circuit or to physically fit in one make of set. Such parts may be available only from the set manufacturer's parts department. Other manufacturers may make some of these parts. There are some parts manufacturers who sell their products through local, independent electronic-parts stores. These companies make exact replacement parts for electronic products if there is enough demand for the parts. The way to find out about the availability of these replacement parts is to check with the local parts-distributor. It is necessary to have make, model number, and original part number information when seeking these replacement parts. Often, if he has no replacement part, the parts distributor will direct the technician to the set manufacturer's parts division.

Parts installation. One of the most frustrating experiences a technician can have is to install a new part, test the set, and find that the set still does not work correctly. There are times when this is because he has been mistaken in thinking that the original part is bad. There are other times when the technician fails to install the replacement part correctly. This could be due to incorrect connection of wires from the set to the part. Possibly two bare wires are touching at the terminals of the part. The possibility of a poor solder connection also exists. Wires need to be clean. The terminal also has to be clean before soldering the connection.

People often think that solder makes an excellent connecting material. This is not true. Solder should be used for making a good *electrical* connection after a good mechanical connection is made at the terminal. In this way, the wire has less chance of breaking loose once the set is returned to the customer.

18-5
Testing the Repair

Let's assume that the bad parts in the set are replaced. What is the next step in the repair process? Some technicians suppose that the repair is done once the part is installed. This is a very poor attitude to have. There is a procedure, which may seem like a ritual, that must be followed after the part is installed. This procedure is called a *checkout,* or *burn-in,* process. This means that the set is played on the workbench, or somewhere in the shop, for a period of time before it is returned to the customer.

The final checkout can be split into two major steps. The first is the reassembly of the set in the cabinet. This is mostly a visual check. It includes checking to see that all plugs and sockets are properly connected. Are all of the wire connections soldered? Are all parts fully connected or plugged into the proper sockets? Are the cables and wires routed correctly? These little things often make big troubles later on if they are not handled correctly during the installation process. It is possible to cut through the insulation on a wire if the wire is pinched between the chassis and the cabinet. Correct routing of wiring minimizes this kind of problem.

The second major check is the *on-the-air* check. This is the time when the set is checked for excessive heat and smoke. Major parts are felt to see if they are operating at temperatures which could be dangerous. The set's controls are operated to ensure that they all work correctly. Any required adjustments are made in order to make sure the set operates as close to its original manner as possible. The cause of the customer's original complaint is checked to see that it is repaired.

In other words, a complete on-the-air check is conducted before the set is returned to the customer. One purpose of this procedure is to promote good customer relations and to cut down on call-back time. A satisfied customer tells friends and relatives about the good work done at the repair shop. Often, using a small amount of glass cleaner on the picture tube or some cabinet cleaner on the set produces a great image for the customer. This little bit of extra effort creates a lot of good will for the repair shop. This type of promotion helps a business grow. Good will is difficult to measure, but it brings great pleasure to both the technician and the customer.

Safety. A very important part of any repair is related to safety. A malfunctioning set is a hazard to its owner. The possibility of a fire exists if the malfunctioning set is operated without being repaired. Some stages in the set have high voltages and large currents in them. These stages are required to have circuit protection devices installed during the manufacturing process. This is done in order to protect the consumer. Resistors, when overheated, may start to burn. Flameproof resistors are required in critical circuits.

These kinds of safety features have to be maintained by the technician. This is for his protection as well as the customer's safety. A visual check of the set is made before it is returned to the customer. Ac line cords are checked for breaks or wear. All safety interlocks are connected properly. The set's back is installed properly so that no one is able to reach in and touch any part. All knobs are installed so that the customer is not able to touch any bare metal parts—this reduces the possibility of an electrical shock. Another check on sets that do not use a power transformer is to see that ac line leakage is below the minimum value for the set. The technical information about the set shows the procedure and allowable values for leakage.

It is wise to check any United States government regulations regarding health and safety standards for the set. There are bulletins published regarding possible health or safety factors for specific sets. These should be checked and any applicable modifications made while the set is in the shop. All of these safety factors are important to the customer. The removal of any potential danger to the customer goes a long way in creating, and maintaining, a good business.

SUMMARY   There is a logical, systematic approach to the repair of any electronic device. The technician applies certain basic rules during the repair procedure. These rules are formulated from the knowledge gained in learning how to repair. The rules, along with technical information about the unit, make the repair process simpler. These rules apply to any kind of electrical or mechanical device. They have to be applied in order to make the repair.

The logical approach in repair work is to first identify the blocks, or areas, of the set that are working correctly. These blocks are eliminated from the total group of blocks in the set. The problem is then isolated in one of the remaining blocks. The next step is to locate a specific stage within the suspected block. Further work locates a specific circuit in the stage. The final step in this process is to locate a specific component. The process goes from large areas of the set down to the small part. This type of procedure is a practical means of repair. Learning to use this type of approach will speed most repair work.

The technician normally is unable to make a repair unless the technical literature for the specific set is available. This material is obtained from the set's manufacturer. It may also be obtained from a local, independent electronic-parts distributor. The technician uses this material to follow signal flow, locate operating voltage test-points, and to obtain procedures for adjustments to the set.

Three general types of test equipment are used in repair work. These types are grouped as signal tracers, signal generators, and voltage measurers. Each type has its own application in servicing. The technician needs to know how and when to use each type of equipment.

Signal tracers are used to follow a signal as it is processed by the set. A known signal is connected at the input of the set. The tracer probe is connected to points in the set in order to locate where the signal stops. When this point is found the trouble is located. It is located between the point at which the last good signal is found and the point where the signal is no longer found.

Two types of signal tracing devices are often used. One is *called* a signal tracer. Its output is a speaker. Signals are traced through the set by means of special probes. The other device is the oscilloscope. Signals are converted into electrical wave shapes. These shapes appear on the face of the CRT. Most technical literature shows sets of typical waveforms found in a specific set. These are located and compared to those observed on the oscilloscope. Major differences in the two patterns indicate a point of trouble.

The second group of test equipment are called signal generators.

These units develop signals which are similar to those found in consumer electronic products. Audio, radio, FM stereo, and TV signals are created in signal generators. The generator may be used to substitute for a lost incoming signal from a previous stage. Often the technician uses a signal generator by starting at the output stages and working back through the set toward the input. When there is no output then the problem area is located.

Voltage testers are the third group of measuring devices. Operating voltage values are found in the technical literature required for effective repair. Voltmeters are used to measure these values. The voltages are then compared to those in the literature. When a difference is found the trouble area is located in the set.

Certain rules for troubleshooting help make the repair work easier. These rules apply to both operating voltages and to signal paths. The technician selects the rule which applies to the specific area of the set under investigation. Use of these rules saves time and makes the repair work simpler.

The technician has to remember that the set is not human. It does not have a mind of its own. Also, the set did work at one time. The technician's job is not to re-engineer it, but rather to return it to its original operating condition.

A system called *bracketing* is used. The brackets are placed around the suspected area, on the schematic. A test is made based upon one of the rules for troubleshooting. The results of the test are used in order to move one of the brackets from its original position. Each test moves the brackets closer together until one specific stage or circuit is left. At this point, other test equipment is often used in order to locate a specific part that has failed.

There are various kinds of rules for troubleshooting. The set of rules presented in this chapter is used with technical literature in order to locate specific trouble areas. The rules for circuit troubleshooting are summarized in Fig. 18-10. They are not in any specific order. The technician selects the proper rules and applies them to the circuits in the set. The results are interpreted and then additional steps are taken in order to reduce the area of suspicion to a specific circuit.

After the defective part is identified, it is replaced with a good part. Some parts are general in nature. These are obtained from local electronic-parts distributors. Other parts are unique to one manufacturer's set, and these often have to be obtained from the set's manufacturer. Many manufacturers have a national parts-distribution system. Special parts are obtained from these outlets. Often, all parts for the repair may be obtained from these outlets.

After the part is obtained, it is installed. Careful installation is a must if the repair is to be correct. Improper installation of replacement parts will often damage the new parts. It certainly does not correct a malfunction in the set when the part is incorrectly installed.

Linear

1. Measure at or near the midpoint of the path.
2. Repeat until the area is reduced to a single block.

Divergent

1. Check one output. If it is normal, then the input is normal.
2. If the output is not normal, test at the common feed point.

Convergent

1. Check each input where it enters the common stage.
2. If both inputs are good, the trouble is in the common stage.

Feedback

1. Open the feedback circuit.
2. An output, with the feedback circuit, open indicates a bad feedback circuit.
3. No output, with the feedback circuit, open indicates a bad forward-going block.

Switching

1. Test outputs by switching to another input.
2. If output is still bad, trouble is in common block.
3. If output is good, trouble is in input block.

Figure 18-10. The rules for troubleshooting circuits.

After repair, the set is checked visually in order to be certain that all connections are made and all wires are properly placed in the set. Final checkout includes operation of the set for a period of time in order to be sure that all repairs are correct, that the customer's complaint is fixed, and that the set works as it should.

Part of the repair process is safety. All checks relative to safe operation of the set are made as a part of the repair process. These checks include review of bulletins published by the set's manufacturer and by the federal government. All necessary modifications are made in order to ensure a safely operating set for the customer.

QUESTIONS

1. What is the most important aspect of troubleshooting? Give reasons for your answers.

2. Describe the process of *bracketing*.

3. What kinds of information are available from the manufacturer's technical literature?

4. What is the difference between a signal generator and a signal tracer?

5. Give examples of where one will find each kind of a signal path in a TV set.

6. State the four major steps of logical troubleshooting.

7. What are two sources of replacement parts?

8. Give reasons for a physical inspection of the set after it is repaired.

9. Give reasons for an on-the-air test after the set is repaired.

10. Why is safety such an important factor in repair work?

In
Conclusion

The intent of this book is to provide a beginning for persons interested in the field of consumer electronic-repair. We begin our study of this subject with a discussion of the devices used to convert energy into electrical waves and of the devices used to convert electrical waves into either visual or audio form. The examples used are devices commonly found in the home or automobile.

The author starts with a consideration of the large block called a radio or a television receiver. As the book progresses, the large block is divided into smaller blocks. Each of the smaller blocks takes on an identity and a purpose. Signals are followed through each receiver used as an example. Kinds of signals are covered.

By this time the student should be doing some exploratory work on actual sets. In class, signals are discussed and followed through the various paths in working sets. These signals and paths are compared to those discussed in this book. The student learns to understand and identify circuits. His study of this book should provide him with a basic understanding of electronics.

This may be the end of the book, but it should be just the beginning of an understanding of electronic theory and application. The reader is encouraged to continue with the quest for knowledge in this field. The door is open for further study of electronics. Many consumer electronic-devices used today were unheard of only a few years ago. Technology continues to change. A person who has a good foundation of theoretical concepts and who is able to apply those concepts as changes occur will succeed in this field.

The challenge is here. It is up to you, the reader, to take it up. Continue in the study of electronics. Build upon the foundation of the material you have learned. The door is open: you, and only you, have to make the move to continue to learn more about this exciting field of electronics.

Appendices

APPENDIX 1
TELEVISION CHANNEL FREQUENCIES

Channel Number	Frequency Band, MHz	Picture Carrier Frequency, MHz	Sound Carrier Frequency, MHz	Channel Number	Frequency Band, MHz	Picture Carrier Frequency, MHz	Sound Carrier Frequency, MHz
2	54–60	55.25	59.75	43	644–650	645.25	649.75
3	60–66	61.25	65.75	44	650–656	651.25	655.75
4	66–72	67.25	71.75	45	656–662	657.25	661.75
5	76–82	77.25	81.75	46	662–668	663.25	667.75
6	82–88	83.25	87.75	47	668–674	669.25	673.75
7	174–180	175.25	179.75	48	674–680	675.25	679.75
8	180–186	181.25	185.75	49	680–686	681.25	685.75
9	186–192	187.25	191.75	50	686–692	687.25	691.75
10	192–198	193.25	197.75	51	692–698	693.25	697.75
11	198–204	199.25	203.75	52	698–704	699.25	703.75
12	204–210	205.25	209.75	53	704–710	705.25	709.75
13	210–216	211.25	215.75	54	710–716	711.25	721.75
14	470–476	471.25	475.75	55	716–722	717.25	721.75
15	476–482	477.25	481.75	56	722–728	723.25	727.75
16	482–488	483.25	487.75	57	728–734	729.25	733.75
17	488–494	489.25	493.75	58	734–740	735.25	739.75
18	494–500	495.25	499.75	59	740–746	741.25	745.75
19	500–506	501.25	505.75	60	746–752	747.25	751.75
20	506–512	507.25	511.75	61	752–758	753.25	757.75
21	512–518	513.25	517.75	62	758–764	759.25	763.75
22	518–524	519.25	523.75	63	764–770	765.25	769.75
23	524–530	525.25	529.75	64	770–776	771.25	775.75
24	530–536	531.25	535.75	65	776–782	777.25	781.75
25	536–542	537.25	541.75	66	782–788	783.25	787.75
26	542–548	543.25	547.75	67	788–794	789.25	793.75
27	548–554	549.25	553.75	68	794–800	795.25	799.75
28	554–560	555.25	559.75	69	800–806	801.25	805.75
29	560–566	561.25	565.75	70	806–812	807.25	811.75
30	566–572	567.25	571.75	71	812–818	813.25	817.75
31	572–578	573.25	577.75	72	818–824	819.25	823.75
32	578–584	579.25	583.75	73	824–830	825.25	829.75
33	584–590	585.25	589.75	74	830–836	831.25	835.75
34	590–596	591.25	595.75	75	836–842	837.25	841.75
35	596–602	597.25	601.75	76	842–848	843.25	847.75
36	602–608	603.25	607.75	77	848–854	849.25	853.75
37	608–614	609.25	613.75	78	854–860	855.25	859.75
38	614–620	615.25	619.75	79	860–866	861.25	865.75
39	620–626	621.25	625.75	80	866–872	867.25	871.75
40	626–632	627.25	631.75	81	872–878	873.25	877.75
41	632–638	633.25	637.75	82	878–884	879.25	883.75
42	638–644	639.25	643.75	83	884–890	885.25	889.75

APPENDIX 2
DECIBEL TABLE

+dB	Voltage Ratio $(R_{in} = R_{out})$	Power Ratio	−dB	Voltage Ratio $(R_{in} = R_{out})$	Power Ratio
0	1.000	1.000	0	1.000	1.000
0.1	1.012	1.023	0.1	0.989	0.977
0.2	1.023	1.047	0.2	0.977	0.955
0.3	1.035	1.072	0.3	0.966	0.933
0.4	1.047	1.096	0.4	0.955	0.913
0.5	1.059	1.122	0.5	0.944	0.891
1.0	1.122	1.259	1.0	0.891	0.794
2.0	1.259	1.585	2.0	0.794	0.631
3.0	1.413	1.995	3.0	0.708	0.501
4.0	1.585	2.512	4.0	0.631	0.398
5.0	1.778	3.162	5.0	0.562	0.316
10	3.162	10.00	10	0.316	0.100
15	5.620	31.60	15	0.178	0.037
20	10.00	100.0	20	0.100	0.010
30	31.60	1000	30	3.16×10^{-2}	10^{-3}
40	100.0	10^{+4}	40	10^{-2}	10^{-4}
50	316.0	10^{+5}	50	3.16×10^{-3}	10^{-5}
100	10^{+5}	10^{+10}	100	10^{-5}	10^{-10}

APPENDIX 3
COLOR CODE FOR WIRING

Color	Abbreviation	Connected to
Red	Red or R	High side of voltage source, $B+$ for tubes
Blue	Blue or B	Amplifier tube plate, transistor collector, FET drain
Green	Grn or G	Tube control grid, transistor base, FET gate, input of diode detector
Yellow	Yel or Y	Tube cathode, transistor emitter, FET source
Orange	Orn or O	Screen grid of tube, second base of transistor
Brown	Brn or N	Heaters or filaments
Black	Blk or K	Chassis ground return
White	Wht or W	Return for control grid (AVC bias), or base of transistor
Gray	Gra or A	AC power line

In addition, blue is used for high side of antenna connections. For stereo connections in audio equipment, the right channel uses red (high side) with green, and the left channel uses white (high side) with blue.

APPENDIX 4

GRAPHIC SYMBOLS

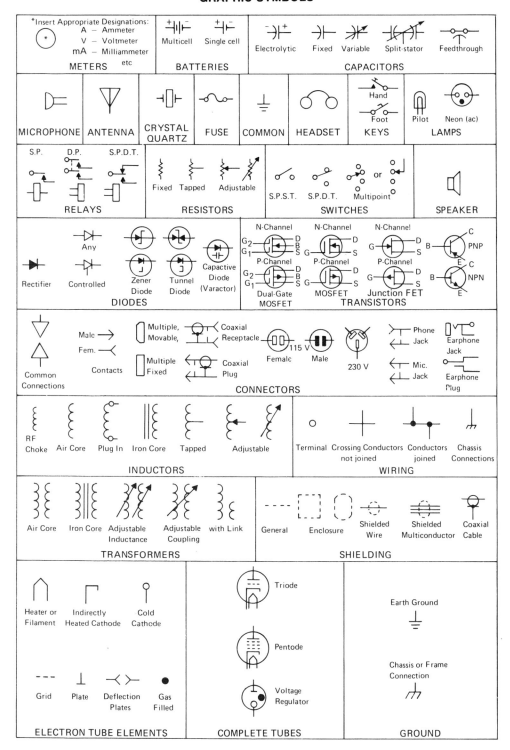

APPENDIX 5

ELECTRONICS INDUSTRIES
RESISTOR COLOR CODE

Color	Meaning of color		
	In Band I or II	In Band III	In Band IV
Black	0	Add no zeros (× 1)	—
Brown	1	Add 1 zero (× 10)	—
Red	2	Add 2 zeros (× 10^2)	—
Orange	3	Add 3 zeros (× 10^3)	—
Yellow	4	Add 4 zeros (× 10^4)	—
Green	5	Add 5 zeros (× 10^5)	—
Blue	6	Add 6 zeros (× 10^6)	—
Violet	7	Add 7 zeros (× 10^7)	—
Gray	8	Add 8 zeros (× 10^8)	—
White	9	Add 9 zeros (× 10^9)	—
Gold	—	Divide by 10 (× 10^{-1})	±5%
Silver	—	Divide by 100 (× 10^{-2})	±10%

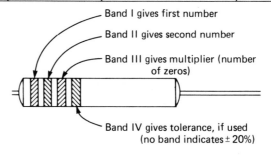

Band I gives first number

Band II gives second number

Band III gives multiplier (number of zeros)

Band IV gives tolerance, if used (no band indicates ± 20%)

Glossary

AC Abbreviation for alternating current.

AGC Automatic gain control. The circuit samples demodulated gain levels and provides an automatic correction bias which maintains a predetermined signal amplitude.

Alternating current Current which has periodic alternations of positive and negative polarities.

Ammeter An instrument that measures the rate of current flow.

Ampere A unit of current.

Amplification The process of increasing the current, voltage, or power of a signal.

Amplifier A device designed to increase the signal voltage, current, or other waveform measured in either a positive or negative direction.

Amplitude The height of the alternating voltage or current measured from zero to its most positive point, or from zero to is most negative point.

Amplitude Modulation The type of modulation commonly used for *standard* radio broadcasting. The *carrier* signal is modulated by low-frequency audio signals so that the overall waveform-amplitude varies above and below the normal carrier level at a rate and amplitude change corresponding to the modulating signal.

Anode The plate, or positive electrode.

Antenna A device used for receiving or transmitting RF signals. Sometimes called an *aerial*.

Attenuate To diminish the amplitude of a signal.

Automatic Frequency Control A circuit designed to stabilize the frequency of an oscillator.

AVC Automatic volume control.

Back Porch The portion (approximately 8 cycles) on the blanking pulse that contains the 3.58-megahertz burst signal.

Band All the frequencies that are within two limits.

Bandpass Amplifier An amplifier designed to amplify the band limits. Term used to define the chrominance amplifier in a color television receiver.

Bandpass Filter An electronic network which passes a specific band of frequencies.

Bias The difference of potential applied between the grid and cathode of a tube or between transistor elements to provide an operating point at zero signal input.

Black Level The base upon which the synchronizing pulses (vertical, horizontal, and 3.58-megahertz burst) rest. This is the voltage level that cuts off the flow of electrons, causing the picture tube's face to blacken.

Blanking Electron beam cutoff in a cathode-ray tube during beam retrace time.

Burst The burst signal originates at the transmitter, and is used in the receiver for synchronizing the phase and frequency of the color oscillator.

Burst Oscillator The color oscillator.

B—Y Signal Difference signal which is a component of the chrominance signal. When added to the brightness, or Y, signal, results in the blue primary-color signal.

Capacitance The quantity of electric charge (usually in fractional farad quantities) which a capacitor is capable of *storing* for a given voltage.

Capacitive Reactance The opposition which a capacitor offers to ac at a specific signal frequency.

Carrier An RF signal capable of being modulated to carry information.

Cathode The element in a vacuum tube which emits electrons.

Cathode Follower A tube circuit in which the output signal appears across the cathode, with the anode at signal ground. Also a grounded-collector transistor circuit.

Cathode-Ray Tube A tube with a phosphor-treated face on which an electron beam traces an image.

Charge The quantity of energy stored by a capacitor or storage-type battery.

Chrominance Amplifier The amplifier section, of a color receiver, that amplifies only the color and burst frequency.

Chrominance Signals Components of a chrominance signal representing hue and saturation of the color information.

Closed Circuit A closed loop through which a current flows when voltage is applied.

Circuit Breaker An electromagnetic or thermal device that opens a circuit when the current exceeds a certain value.

Class A Amplifier An amplifier biased to operate on the linear portion of the characteristic curve.

Class B Amplifier An amplifier biased to operate at or near the tube or transistor cutoff point. Positive alternations of the input signal cause current flow.

Class C Amplifier An amplifier biased beyond the cutoff point so that current flows for only a portion of the positive alternations of the input signal.

Cold-Cathode Tube A tube which requires no external heat-current source to produce electron emission.

Color Killer Circuit that cuts off the color amplifier during the time monochrome signals are being received. This prevents color interference in the black-and-white picture.

Color Oscillator Oscillator frequency at 3.58 megahertz. Its purpose is to furnish the continuous subcarrier frequency required for demodulators in the receivers.

Compatability Permits use of a black-and-white receiver to reproduce color pictures on a shade scale from black to white, and permits the color receiver to reproduce black-and-white transmitted signals.

Composite Color Signal All of the color signals transmitted by amplitude modulation. Includes the luminance signal, chrominance signal, sidebands, burst signal, horizontal and vertical, and equalizing pulses, together with the blanking pulse.

Conductance The current-carrying ability of a wire. The unit value is *mho*. It is the reciprocal of resistance.

Conductor A medium which carries a flow of electricity.

Continuous Wave An unmodulated RF waveform of constant amplitude. Usually the term applied to a wave transmitted in bursts of short and long duration to form the Morse code.

Control Grid The grid to which a signal is usually applied in a tube.

Converter The stage in a superheterodyne receiver which produces the IF signal by mixing the RF carrier with a locally generated signal.

Coupling The effective *linkage* connecting two electronic circuits—usually transformers, capacitors, and inductors.

Crystal Oscillator A signal generating circuit in which the frequency is controlled by a piezo-quartz crystal.

Cutoff Frequency The frequency of a filter or other circuit beyond which signal flow ceases.

Cycle In ac, one complete alternation of positive and negative.

Dc Amplifier An amplifier using direct coupling (no coupling capacitors or transformers).

Decibel A unit for expressing the ratio of two amounts of electric power.

Deflection Bending of the electron beam, both in the horizontal and vertical direction. The horizontal deflection performs the function of tracing the picture lines while the vertical deflection allows formation of fields and frames.

Delayed AVC An automatic volume control circuit designed to produce an AVC bias only for signals above a fixed amplitude.

Demodulation A signal rectifying system which extracts the modulating signal component from the modulated carrier.

Detection To separate modulation from the signal.

Dielectric The insulating material between the two conductors, such as in a capacitor, or the insulating material between transmission line conductors.

Diode A two-element tube or two-terminal solid-state rectifier.

Direct Current Current flow in one direction.

Discriminator The *detector* used in frequency modulation. It is used also to compare two ac signals.

Distortion Unwanted modification of desired signal.

Doubler A circuit in transmitting systems which doubles the frequency of the input signal. In power supply systems, a circuit for doubling voltage amplitude.

Electrode A terminal used to emit, collect, or control electrons.

Electrolyte A solution or a substance which is capable of conducting electricity; it may be in the form of either liquid or paste.

Electrolytic Capacitor A capacitor utilizing an electrolyte to form the dielectric insulation.

Electromagnet A magnet made by passing current through a coil of wire wound on a soft iron core.

Emitter Transistor electrode similar, functionally, to the cathode of a tube.

Farad The unit of capacitance. Fractional values are used in practical electronics.

Feedback A transfer of energy from the output of a circuit back to its input.

Feedback Oscillator A signal-generating circuit that employs regenerative feedback to sustain oscillations.

Ferrite A metallic compound used for high Q core-materials in inductors.

Field Every alternate horizontal line of television picture, as scanned during $1/60$ second. One field begins with a full line from the upper left-hand corner, the next with a half line beginning midway across the top. The two fields take a time of $1/30$ second, and form one frame. There are 525 lines per frame, and $262 1/2$ lines per field.

Filament The electrode in a vacuum tube which is heated for electron emission or which transfers its heat to a separate cathode.

Filter A circuit designed to pass certain signal components and attenuate others.

Filter Capacitor An electrolytic capacitor used in power supplies to reduce ripple.

Filter Choke An inductor used in power supplies to reduce ripple.

Forward Bias The bias applied between the base and emitter of a transistor.

Frequency The number of complete cycles per second in an alternating wave. A cycle includes negative and positive *excursions*.

Frequency Modulation A system where the frequency of the carrier signal is shifted above and below its normal *center* frequency by the modulating signal.

Frequency Response A graph or curve depicting the relative gains of an amplifier at all frequencies.

Fullwave Rectifier A power supply circuit which uses both alternations of the ac waveform to produce direct current.

Gain The ratio of the output to the input signal, voltage, or current.

Generator A machine that changes mechanical energy into electrical energy by rotating coils of wire within a fixed magnetic field.

Grid A wire, usually in the form of a spiral, used to control the electron flow in a vacuum tube.

G–Y Signal A green-minus-Y signal is the color difference signal for green. When combined with the proper proportion of Y, it results in the green color.

Halfwave Rectifier A tube or solid-state diode that converts ac to pulsating dc by rectifying one alternation of each ac cycle.

Harmonic A signal related to a fundamental signal by some integral multiple.

Heater A vacuum tube electrode which heats the cathode.

Henry The basic unit of inductance. One henry represents the amount of inductance present when a current change of 1 ampere per second produces an induced voltage of 1 volt.

Heterodyne The electronic mixing of two signals of different frequencies to produce a third signal.

High-Pass Filter A circuit that transfers high frequency signals while attenuating the low frequencies.

Horizontal Sync Timing of the receiver's horizontal oscillator so that it coincides with the master oscillator at the transmitter.

Impedance A combination of resistance and reactance which opposes ac current flow.

Inductance The property, of a coil, which opposes a change in current.

Inductive Coupling Magnetic lines of force produced by the flow of current in one coil *coupled* to another coil and produce a flow of current.

Inductive Reactance The opposition an inductor offers to ac for a given signal frequency. It is measured in ohms.

In Phase The condition that exists when two ac waves of the same frequency pass through their maximum and minimum values of like polarity at the same instant in time.

Interlaced Scanning Every other line of the image is scanned during one downward sweep of the scanning beam, and the remaining lines are scanned during the next downward sweep of the scanning beam.

Intermediate Frequency The signal obtained by heterodyning or mixing two signals of different frequencies.

Kilo A prefix meaning *1,000.*

Limiter A circuit which limits the peak amlitudes of signal waveforms to a predetermined level.

Load A resistor or transformer, usually, across which the output signal of a tube or transistor is developed.

Low-Pass Filter A circuit designed to pass low frequency signals and attenuate the high frequencies.

Luminance Same general meaning as brightness when refering to color television.

Luminance Channel The channel intended primarily for carrying luminance information.

Luminance Signal Controls brightness, but not color. Has a frequency range of 4.0 megahertz, for reproduction of fine detail in the picture. By itself, this signal can produce a monochrome picture.

Magnetic Field The area in which magnetic lines of force exist.

Matrix A group of resistors through which the luminance signal combined with signals from the demodulators form the color primary signals and the $G-Y$ color difference signal at the transmitter.

Mho The unit of conductance, transconductance, or admittance. The word *ohm* spelled backwards.

Micro A prefix meaning one millionth.

Milliammeter An ammeter constructed to measure fractional (thousandths) values of an ampere.

Modulation The process of modifying an RF carrier signal to transmit audio or video signal information over great distances.

Monochrome Description of a picture that appears in black, white, and shades of gray, but has no coloring.

Ohm The unit of resistance.

Oscillator A regenerative circuit designed to produce signals.

Oscilloscope An instrument using a cathode-ray tube which presents a visual display of electric signals or waveforms.

Parallel Resonant Circuit A circuit that is tuned by the use of a coil in parallel with a capacitor. At resonance, it offers a high impedance path to the current flow. This permits a large value of signal voltage to appear across this circuit.

Peak-to-Peak Value The overall amplitude of a signal, measured from its lowest (or most negative) peak to its highest (or most positive) peak.

Phase Inverter A circuit that changes the phase of the voltage or current by 180 degrees.

Phosphor A fluorescent material used for the screen in a cathode-ray tube.

Picture Element The smallest portion of a picture, or scene, that is converted into an electrical signal.

Potential A voltage that, compared to a reference point, is more positive or more negative.

Potentiometer A variable resistor.

Power Amplifier An audio or RF amplifier designed to deliver signal energy (power) rather than signal voltage.

Power Supply A circuit designed to furnish operating voltages and currents for electronic devices.

Preamplifier An additional stage of amplification preceding another amplifier to increase signal amplitudes above a given level.

Push-Pull Circuit Push-pull normally refers to an amplifier circuit with two tubes or transistors operating so that when one is conducting on a positive alternation, the other operates on a negative alternation.

Radio-Frequency Amplifier An amplifier designed to increase RF signal levels.

Raster The phosphor emission of light when the electron beam strikes the surface. The beam is deflected both horizontally and vertically, and no picture information is present.

Ratio Detector A dual-diode frequency-modulation *detector.*

Reactance The opposition to ac current offered by an inductor or capacitor.

Resonant Frequency The frequency which produces resonance in a coil-capacitor tuning circuit.

Rectifier A device that changes alternating current to unidirectional current.

Regulation The degree to which voltage is held near its no-load value when a load is applied.

Resistance The opposition to current flow. It's measured in ohms.

Resonance A condition in a tuned circuit in which reactances cancel at a specific frequency.

Retrace Blanking Voltage pulses for blanking are derived from the vertical sweep oscillator or the deflection circuit. These pulses are used to darken the picture tube during retrace intervals.

R−Y Signal A red-minus-Y, or red-minus-luminance (brightness), signal. When combined with the plus-luminance (+Y) signal, it results in the red primary-color signal.

Saturation The point in a tube or transistor at which gain levels off despite further attempts to increase it.

Scanning The process by which each picture element is reproduced on the screen of the picture tube.

Shadow Mask A thin sheet containing as many small openings as there are groups (triads) of phosphor dots. It is positioned directly behind the screen.

Silicon Controlled Rectifier A solid-state rectifier in which conduction can be started by applying a control voltage.

Sinusoidal Having the form of a sine wave.

Solenoid An electromagnetic coil with a moveable plunger.

Subcarrier In color television, a signal with a frequency of 3.58 megahertz. The chrominance signals modulate this frequency.

Sync Pulse In composite television signal, this pulse rides on top of the blanking pulse. It is used to synchronize the vertical-receiver and horizontal-receiver oscillators.

Thermistor A resistor that changes its resistance value to compensate for temperature changes.

Trace A visible line on the screen of a cathode-ray tube.

Transducer A device for converting energy from one form to another, such as vibrations, from a phonograph pickup, into audible sounds.

Transformer A device with two or more coils linked by magnetic lines of force. It's used to transfer energy from one circuit to another.

Tuned Circuit A circuit at resonance.

Unmodulated An RF carrier signal with no modulation.

Video Amplifier A circuit capable of amplifying a very wide range of frequencies, from the audio band and higher.

Volt The unit of electrical potential (EMF).

Voltage Divider Resistors placed in series across a voltage to obtain intermediate values of voltage.

Voltage Doubler A power supply circuit so designed that the rectified voltage amplitude is almost double the input ac amplitude.

Watt The unit of electric energy or power.

Waveform The shape of the wave obtained when instantaneous values of an ac quantity are plotted against time in rectangular coordinates.

Yoke In a television receiver, a coil arrangement around the neck of the picture tube which provides electromagnetic deflection of the CRT beam vertically and horizontally.

Y-Signal The luminance, or brightness, signal in color television. This signal contains the high-definition details.

Zener Diode A solid-state semiconductor that has voltage regulation characteristics when subjected to reverse bias.

Index